Hi-Tech Horticulture

Volume 1: Crop Improvement
Nursery and Rootstock Management

OTHER VOLUMES OF THE SERIES

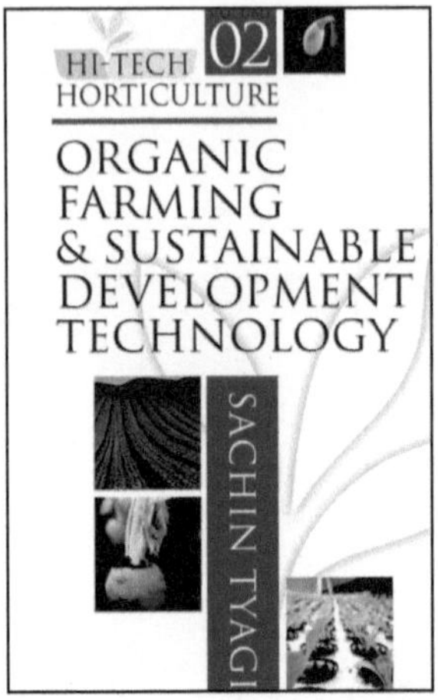

ISBN 9789387973411
INR 2700.00

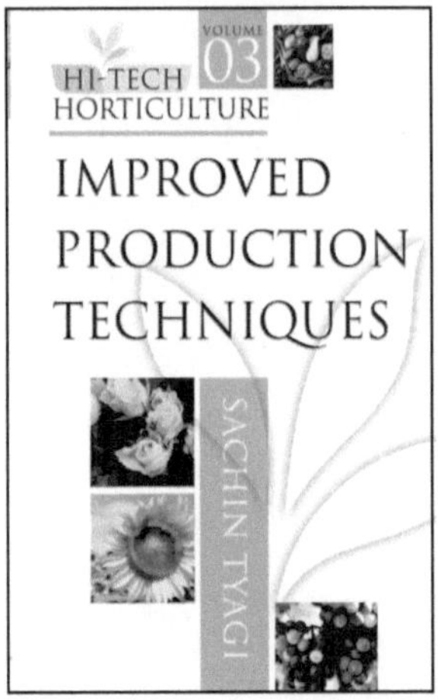

ISBN 9789387973428
INR 2700.00

ISBN 9789387973435
INR 2700.00

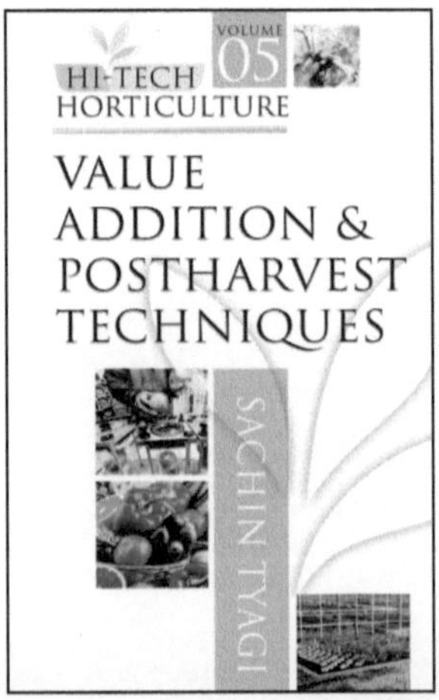

ISBN 9789387973459
INR 2700.00

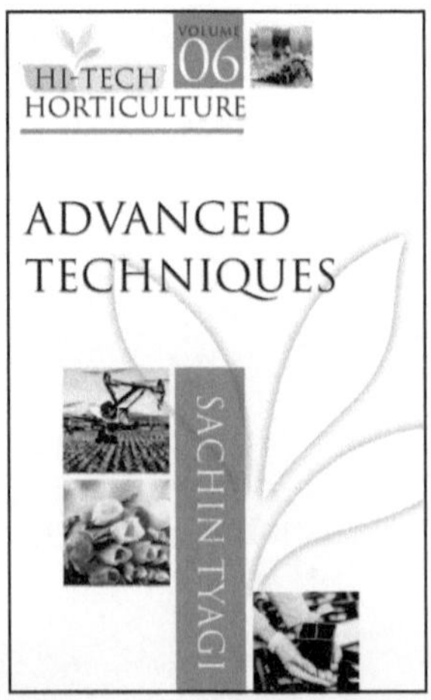

ISBN 978-9387973466
INR 2700.00

Hi-Tech Horticulture

Volume 1: Crop Improvement Nursery and Rootstock Management

Sachin Tyagi
Department of Horticulture (F & FT)
Bihar Agriculture University
Sabour, Bhagalpur-813210 (Bihar) India

NEW INDIA PUBLISHING AGENCY
New Delhi – 110 034

NEW INDIA PUBLISHING AGENCY
101, Vikas Surya Plaza, CU Block, LSC Market
Pitam Pura, New Delhi 110 034, India
Phone: + 91 (11) 27 34 17 17 Fax: + 91 (11) 27 34 16 16
Email: info@nipabooks.com
Web: www.nipabooks.com

Feedback at feedbacks@nipabooks.com

ISBN: 978-93-87973-40-4

Composed and Designed by NIPA

Preface to the Series

Horticulture plays an important part in today's agriculture and there are new avenues that are being achieved by horticulture. The subject has transformed from only being about vegetables, fruits, flowers and postharvest technology and has moved towards disease, breeding, pathology, physiology, greenhouse technologies and other areas which were never heard for.

The book series: Hi-Tech Horticulture has been worked out keeping the above mentioned issues in mind with contribution by eminent professors and scientists. The volumes are on:

1. Crop Improvement Nursery and Rootstock Management
2. Organic Farming and Sustainable Development Techniques
3. Improved Production Techniques
4. Plant Protection and Stress Management
5. Value Addition and Post-Harvest Management
6. Advance Techniques

Series Editor

Preface to the Volume

The present volume has been contributed by eminent professors and scientist in their field and updated and most recent developments have been covered with.

I acknowledge the support from my university colleagues and contributors.

Encouragements by the present Vice Chancellor Dr Ajoy Kumar Singh is duly acknowledged.

Sachin Tyagi

Contents

Section 03: Nursery Management

Section 04: Rootstock Management

Section 01
Tools and Techniques of Crop Improvement

1

Breeding for Disease Resistance A Journey from Conventional to Molecular Approach

Rishav Kumar[1] and Sachin Tyagi[2]

[1]Department of Plant Breeding and Genetics
Bihar Agricultural University Sabour, Bhagalpur-813210, India
[2]Department of Horticulture (Fruit and Fruit Technology)
Bihar Agricultural University, Sabour, Bhagalpur -813210, India

Abstract

Modern agriculture has disturbed the age-old balance of host-pathogen system resulting diseases occur at regular interval leading to enormous loss to the agricultural production, which is estimated to be 10-25% or less during normal years but may extend up to 100 % during epidemic. Broadly, there are two breeding approaches adopted for disease resistance which includes conventional and molecular one. Conventional approach includes pureline, pedigree, backcross, recurrent selection and interspecific hybrids through which various resistant varieties has been developed. However, after sometime there is sudden outbreak of diseases due to continuous sowing of these varieties which results in the advancement in molecular biology and with the discovery of trait linked markers it paved a new pathway to look a new corner to develop varieties which are disease resistant. Molecular approach consists of gene pyramiding through which two or more genes providing resistance to different types of diseases is introgressed in a particular variety, gene deployment and multiline varieties which consists of heterogeneous population formed by mixing the cultivars to differ in the ability of disease resistance and reduces the rate of epidemic development. With the help of makers in marker assisted breeding using foreground, recombinant and background selection several linked markers are identified which provides resistance against disease like in wheat stem rust, leaf rust etc. With the development of more advanced technologies transgenic strategies came into existence where gene silencing is done through RNA mediated silencing techniques and several fungal, viral and bacterial diseases has been controlled. Apart from that forward breeding and reverse breeding also plays a significant role in controlling diseases by manipulating various defence mechanism. Last but not the least is the significant role of tissue culture in controlling the disease. Various toxin and culture filtrates have been developed which provides resistance against several diseases. Through somaclonal variations sugarcane variety like Ono has been released resistant against Fiji diseases. Hence we can say that breeding for disease resistance plays a significant role in controlling some of the diseases and has travelled a long journey from conventional to the molecular one.

Keywords: *conventional approach, molecular approach, forward breeding, reverse breeding*

Introduction

Diseases are one of the most dangerous enemies of crop which leads to entire failure of crop plant. From ancient times it occurs in minute form to most severe form leading to complete failure of crop. Some of them become epidemic also like Bengal famine, late blight of potato, powdery mildew of grapes and coffee rust. On these crops not only the livelihood of farmers depends but it also affects the economy of the nation whose GDP depends on it and its failure leads to breakdown of economy which is backbone of any nation for prosperity of its citizens. If proper management doesn't takes place at appropriate time then it becomes havoc and may bring a prosperous country on the brink of poor country. The first reported work regarding disease resistance is of Theophrastus during the 3rd B.C. who cultivated varieties differed in the ability to avoid disease. Later, de Barry 1861 first established the parasitism of stem rust of wheat. Erikson in 1894 showed that pathogen morphologically similar differed from each other in their ability to attack different related host species. But the major milestone came with the work of Flor in 1956 who gave gene for gene hypothesis. With the time several researchers also came and focused on development of high yielding variety with adequate grain quality but to permit high yield diseases must be controlled.As we know that when a variety is released its area increases and covers large area but as a consequence if its new virulent race of pathogen also develops and is the bust period for crop hence cultivar failure occurs and leading to heavy economic loss which opens a broad scope for the plant breeder to search for the new technique to devise regarding this field. One of the effective way to control them is use of resistant varieties but before using any strategy first of all one must know at which level diseases are causing economic losses and secondly, by application of which measures they can be controlled. Some of them have been controlled by removal of its alternate host, volunteer plant or through, chemical biological and cultural practices like bacterial blight of rice but effective control has not been obtained (Mishra *et al.,* 2016). Common bunt and loose smut has been controlled by use of effective fungicides. But as we know that continuous use of these chemicals not only pollutes the environment but also makes responsible for the appearance of new virulent races of pathogen which is only not harmful to human beings but also to the other flora and fauna. Hence by considering the above genuine reasons it is necessary to look forward towards every nook and corner which not only enlightens the lamp for its effective control but also opens a broad spectra in new area for getting better yield without affecting the natural resources for the benefit of human beings as well as with the harmony with nature. Hence, breeding for disease resistance came into existence which covers a long distance journey from conventional to the molecular methods.

A) Conventional approach

It is an age old collection of methods which has been developed from ancient times. From beginning of human civilization in society form till today various cultivars has been developed. The most renowned examples is development of semi-dwarf high yielding cultivars of cereals during the Green Revolution and the hybrid rice developed in 1970s.Some of them are: -

(i) **Pure line method** which is developed especially for the self-pollinated crops. Since pureline cultivar is progeny of single, homozygous, self-pollinated species so they can be easily developed in nature and has importance in breeding of inbred lines that are used to make hybrids in self or cross pollinated crops as well as they are desired in regions where uniformity of a product fetches a high premium. Through this method maximum possible improvement over the original varieties can be achieved and useful in improving low heritability traits since selection is based on progeny performance. Some of the pure line cultivars are Suvita-2 of cowpea which is resistant against striga is selection from landrace in Burkina Faso. Kanred variety of wheat is resistant against rust is selection from Kansas from Crimean variety, an introduction from Russia. Other varieties are Pusa Sawani of okra resistance for yellow mosaic virus, dwarf yellow milo of sorghum resistance for red rot, Buri 107 of cotton resistant for wilt. But it requires more time for release of any varieties than mass selection, more susceptible to diseases, has poor adaptability and stability in production. In case of developing a new variety in cross pollinated crops it is expensive as well as laborious.

(ii) **Pedigree method** in which the detailed record of the origin of the selected individuals or lines are maintained. Due to such records we are able to know from which parent superior F_2 progeny will develop. It is useful for highly heritable traits like seed size, shape, disease, insect, height and maturity. It is widely used for development of disease resistant variety if governed by major genes (Miah *et al.*, 2013). Variety developed through pedigree method is Laxmi of cotton resistance for leaf blight. But the varieties developed through it cannot be used in environments where genetic variability for the characters of interest is not expressed, requires a lot of time for testing along with intensive record keeping. Also it is costly in terms of land and labour requirement. In case of characters governed by polygenes it is not effective.

(iii) With the time backcross variety has been developed by crossing F_1 hybrid with any of the known parent and is useful in breeding for small grains crops. In this method two plants are selected and crossed together. On regular backcrossing with one of the recurrent parent the backcross progeny developed is almost identical to the recurrent one. In this method the desirable variety which are lacking in some characteristics known as a recurrent or recipient parent, while the undesirable variety or wild variety has only one or two desirable characters. This method is practiced in South and Southeast Asia (Joseph *et al.*, 2004; Toojinda *et al.*, 2005) as breeding strategy to improve elite varieties such as KDML105, Basmati and Manawthukha for resistances to blast (Sreewongchai *et al.*, 2010). Advantages of using this method is intervarietal transfer of characters(disease resistance, seed colour, plant height,earliness, seed size, seed shape), interspecific transfer of disease resistance from related species to cultivated species, transfer of cytoplasm from one variety or species to another which is desirable in case of cytoplasmic male sterility, development of transgressive segregants and production of isogenic line. Some of the varieties which has been developed through this method are BD 8 of cotton resistant for wilt, MS 521, MS 541A, MS 570 A of bajra resistant for downy mildew, transfer of wilt resistance to alfalfa variety California Common from the variety Turkestan. But variety developed through it requires many backcrosses to produce a new cultivar, the new variety generally cannot be superior to the recurrent parent, except for the character that is transferred, hybridization has to be done for each backcross which makes it often difficult since it is time taking as well as costly and by the time the backcross is over, the recurrent parent may have been replaced by other varieties superior in yielding ability and other characteristics.

Also this method is poor in quantitative traits which is more restricted for recessive traits, sections of genome from the non-recurrent parents are often still present and has unwanted traits associated with them. For very wide crosses, limited recombination may maintain thousands of 'alien' genes within the elite cultivar.

(iv) **Recurrent selection** is being also used from ancient times for development of male sterile lines but now a days some of the disease resistant varieties like CG-91 has been also developed against blast (Guimaraes and Correa-Victoria 2000; Courtois *et al.*, 1997). Apart from that, phenotypic recurrent selection is being also used in case of rose for developing varieties resistant against black spot (*Diplocarpon rosae*), rust (*Phragmidium* spp.), cercospora (*Cercospora puderi* and *Cercospora rosicola*) and anthracnose (*Sphaceloma rosarium*). (Debener and Byrne 2014).

(v) **Interspecific hybrid** develop by crossing two species of the same genus with the objective of transferring one or few simply inherited characters like disease resistance to a crop species.Through introgression genetic variation of one or both of the parents increases. Variety developed through interspecific hybridization in potato is Kufri jyoti which is resistant to late blight. Through interspecific hybridization gene for resistance to cotton rust caused by *Puccinia cacabata*has been transferred from *Gossypium anomalum* and *Gossypium arboreum* into *Gossypium hirusutum*(Anjum *et al*., 2015). Aminu, 1940 reported the combination of genes for cotton leaf curl virus resistance and other diseases between *Gossypium hirusutum* and *Gossypium arboretum.* Similarly, introgressed resistance gene B6 found in 'A' genome of *Gossypium arboreum* against bacterial blight caused by *Xanthomonas malvacearum* into *Gossypium barbadense* (Knight, 1957 and Brinkerhoff, 1970). Asource of common bunchy top plant resistance in *Gossypium hirsutum* has been identified in 'Delta Opal', which is used to transfer disease resistance from this variety (Ellis *et al*., 2016). Interspecies gene transfer is a very powerful tool to transfer resistance between species and can there by results in a broader resistance spectrum. In case of wheat *Lr34res* allele has been identified which gives resistance in maize against rust pathogen and the hemibiotrophic fungus that causes northern corn leaf blight disease (Sucher *et al*., 2016). *Lr34res* has already been shown to be effective against the hemibiotrophic fungus rice blast in rice (Krattinger *et al*., 2016) and various biotrophic fungi such as rust and powdery mildew pathogens in wheat and barley

(Risk *et al.*, 2012; Risk *et al.*, 2013). Furthermore, *Lr34res* has been associated with resistance against the hemibiotrophic fungus spot blotch (*Bipolaros sorokiniana*) in wheat (Lillemo *et al.*, 2013). Qualitative and quantitative genes have been identified in maize, for race specific *Ht1* gene (Welz and Geiger, 2000) and *Htn1* that confers resistance against most prevalent NCLB races (Hurni *et al.*, 2015).

(vi) **Mutation breeding** is used as complementary to conventional breeding, since this technique is very effective for imparting disease resistance. Mutagens include physical mutagens (a) Ionizing radiations which is classified into particulate radiations (alpha rays, beta rays,fast and thermal neutrons) and non-particulate radiations (X-rays, gamma rays) (b) Non-ionizing radiations (UV rays). Chemical mutagens include alkylating agents (Mustard or sulphur mustard, nitrogen mustard, ethyl methane sulphonate, methyl methane sulphonate, ethylethane sulphonate, N-methyl-N'-nitro-N-nitrosogunanidine), base analogues (5-Bromouracil, 2-Aminopurine), acridine dyes (acriflavin, proflavin, acridine orange, ethidium bromide), deamination agents (nitrous acid) and other chemical mutagens (hydroxylamine, sodium azide, DNA sequences).Out of the given mutagens fast neutrons, thermal neutrons, X rays, EMS and sodium azides are commonly used in plants but now a days gamma rays is most widely used for induction of mutation in crop plants (Singh 2013). Bhabha Atomic Research Centre (BARC) has developed 41 new crop varieties through radiation treatment which is notified by Ministry of Agriculture, Government of India for commercial cultivation. Some of the varieties are listed below.

Crop	Variety	Release	Released for	Remarks
Soybean	TAMS 98-21	2007	Maharashtra	Resistant to bacterial pustules, Myrothecium leaf spot and soybean mosaic virus diseases
	TAMS-38	2005	Maharashtra	Resistant to bacterial pustules, Myrothecium leaf spot
Brassica juncea	TPM-1	2007	Maharashtra	Tolerant to powdery mildew
Green gram	TM-96-2	2007	Andhra Pradesh	Resistant to powdery mildew and Corynespora leaf spot
	TJM-3	2007	Madhya Pradesh	Resistant to powdery mildew, Yellow mosaic virus and Rhizoctonia root rot
	TMB-37	2005	Eastern UP, Bihar,	Tolerant to yellow mosaic virus

			Jharkhand, Assam, West Bengal	
	TARM-18	1995	Maharashtra	Resistant to powdery mildew
	TARM-1	1995	Maharashtra Gujarat, MP AP, Kerala Karnataka,Tamil Nadu,Orissa	Resistant to powdery mildew
	TARM-2	1992	Maharashtra	Resistant to powdery mildew
	TAP-7	1983	Maharashtra, Karnataka	Tolerant to powdery mildew
Black gram	TU 94-2	1999	Andhra Pradesh, Karnataka, Kerala, Tamil Nadu	Resistant to yellow mosaic virus

Source : http://www.iancas.org/oct07.pdf

Through this method new gene alleles are created which is not present in germplasm pools, induction of new gene alleles into the new varieties that can be used directly as a commercial variety. But having such importance there are some lacuna which restricts the use of mutation breeding in day to day life as there is limited power in generating the desired dominant alleles and is also less effective than cross breeding for a trait which needs for a combination of multiple alleles (Miah *et al*., 2013). Several mutant lines has been developed such as Mahsuri Mutant, SPM 129, SPM 130 and SPM 142 (Azlan *et al*., 2004) for blast resistance in Malaysia (Mohamad *et al*., 2006). In China, the mutant rice variety 'Zhefu 802' deriving from variety 'Simei No. 2' induced by gamma rays has high resistance to rice blast (Shu *et al*., 1997; Ahloowalia *et al*., 2004).

B) Molecular approach

With the discovery of molecular markers this approach came into existence. The development of gene identification technologies using the tools of biotechnology provides ample opportunities for scientists to further improve modern cultivars. Generally breeders use major gene resistance for developing a new variety but with the passage of time it fails to provide the long term disease control which leads to epidemic (Gnanamanickam *et al*., 1999). Molecular markers for trait selection has numerous advantages over morphological markers used in conventional plant breeding. Through this diverse sources of resistance to major diseases with multiple resistance to diseases has been developed. Recent advances in cellular and molecular biology has provided tools to develop more durable resistant cultivars and to overcome the problem of lack of donors for resistance to some diseases (Khush, 2005). Some of the strategies which is in general used for durability of resistance in a cultivar is gene pyramiding, gene deployment, and marker assisted breeding.

(i) Gene Pyramiding or gene stacking is a new concept that came with the breakdown of monogenic resistance where incorporation of two or more major genes in host for specific resistance to the pathogen. This approach has been used in oat against crown rust in Canada. Through gene pyramiding horizontal resistance can be effectively exploited. One of the famous epidemic disease which has occurred and created havoc to the economy like Irish famine caused by *Phytopthora infestans* leads to complete failure of crop. Hence, it was felt that if pyramiding several genes for disease resistance in a single cultivar then it may provide a long term resistance resulting in reduction of rate of reproduction of the pathogen. Using regular backcrossing for the development of genes stacked variety takes a long period of time thereforediscovery of markers makes the gene pyramiding somehow easier. Through gene pyramiding some of the bacterial blight resistance cultivars has been developed *viz.*, IR 64 with *xa 5, Xa 7, Xa 21* at CRIFC Indonesia, IR 64 with *xa 4, Xa 5, Xa 7, Xa 21* at Philippines, Samba Mahsuri with *xa 5, Xa 13, Xa 21* at DRR India, Swarna and IR 64 with *Xa 4, xa 5, xa 13, Xa 21* at CRRI India. Apart from that some of the bacterial blight resistant lines which have been developed in India are IR 24, Lalat and Tapaswani. Major problemsassociated with gene stacking is for finding resistance gene. However,if anyhow it is found then there is problem of linkage drag and many times it is also seen that there is adverse effect on yield of the crop.

(ii) Gene Deployment: It is the planned use of major genes in development of resistant cultivars for various geographical areas. It was first used for soybean breeding in Northern United States. Variety of resistance genes have been identified and characterized. Together, with the development of efficient plant transformation systems these genes offer an alternative means to introduce specific resistance to crop improvement programme. Heterogeneous population is formed by mixing the cultivars which has ability to differ in the ability of disease resistance and reduces the rate of epidemic development (Borlaug, 1958; Browning and Frey, 1969; Jensen 1952; Wolfe *et al*., 1983). Strategy involved in this regard is use of distinct type of resistance mechanism in each variety and use of varieties in a predetermined pattern (temporal or spatial). Based on crop cultivation practices, seasonal and regional preferences for different location specific varieties are used. As a result, distinct varieties has been developed using diverse sources of resistance. Even among varieties used for a particular season, variety with different maturity period should consist of distinct sources of disease resistance. This situation will slow down the development of new virulent races and improve the durability of resistance in present varieties. Among many strategies, distinct gene deployment in different maturity groups may help to improve the durability of resistance in newly developing varieties. It helps in minimizing selection pressure and thereby preventing evolution of resistance in the pathogen population (Bonman, 1992). But there are two difficulties associated with it. First one is that there is no surety for durable resistance and to the second one is due to epistatic gene interaction it is not necessarily possible to select for the presence of multiple genes or for specific gene combinations. To overcome this molecular marker plays a significant role by improving the efficiency as well as resolution of genetic analysis, particularly when multiple resistance genes are exists in a single cultivar (Paterson, 1991). Secondly, markers linked to resistance genes provide tools to facilitate the selection of lines carrying resistance genes in desirable combinations (Abenes, 1993). According to Gould, 1983 gene deployment strategies can be broadly divided into two categories i.e. spatial and temporal.

a) **Spatial gene deployment:** In this gene deployment is done over small geographical region like at the farm level, field level, regional level which includes variety mixture, species mixture, multilines and multiblends (mixture of existing lines or varieties). Variety mixture includes homogenous spatial mixture of different genotypes of one plant species in a field (Adugna, 2004). Resistance in a mixture can be obtained through decrease in inoculum potential, cross protection (Burdon, 1993; Wolfe, 1985). According to Jorgensen, 1993 when a mixture of cultivar having different gene for disease resistance is used it provides a long term durable resistance. Oat and barley mixtures are used for controlling oat sterile dwarf virus.

b) **Temporal gene deployment:** It can be achieved through two approaches either through variety resistance having different genes for disease resistance or by sequential release of variety having genes resistance for particular disease and regularly replacing with new variety with the appearance of new disease having different gene for disease resistance.

(iii) Multiline variety: It is the mixture of several similar purelines having different genes for disease resistance. For development of multiline varieties several purelines with different resistance genes should be developed through back cross programmes using one recurrent parent which is done by transferring disease resistance genes from several donor parents carrying different resistant genes to a single recurrent parent. Each donor parent is used in a separate backcross programme so that each line has different resistant gene. According to Borlaug, 1958 a total of eight to ten lines are used which require eight to ten backcrossing for development of multiline variety which helps in exploitation of horizontal resistance. The durability of resistance depends upon the rate of disease development, number of lines components in a mixture and the extent of planted area. According to Marshal, 1977 two approaches have been proposed for multiline varieties i.e. clean crop approach and dirty crop approach. According to clean crop approach, each donor has different genes for resistance to the given disease so that each pureline obtained from backcross progenies has a different resistance genes. If any component line becomes susceptible to the pathogen it is immediately replaced by a new resistant line. According to dirty crop approach, a multiline variety consists of several lines none of which is resistant to all the known races of the pathogen. Such a multiline variety is regularly attacked by pathogens but the spread of disease is expected to be much slower than in a pureline variety. Varieties of wheat developed through multiline are KSML 3 (consists of 8 lines having rust resistance genes from Robin, Ghanate, Kl, Rend, Gabato, Blue Brid, Tobari), KSML 7406 (mixture of 9 lines), MLKS 11 (mixture of 8 lines derived from E 6254, E 6056, E 5868, Frecor, HS 19, E 4894 etc.), Sonalika Multiline 1, Tumult, Miramar varieties of wheat. Apart from that E 68, E 70, M 68, M 69 varieties of oat has been also developed. Variety developed through multiline has an advantage like all the lines are identical to the recurrent parent in quality and only one or few lines of mixture becomes susceptible to the pathogen in one season. Major shortcoming associated with multiline variety isfrom farmer's point of view as it is difficult for them to change the seeds on every few years depending upon the change in the races of pathogen and by the time a multiline is developed its recurrent parent is immediately replaced by the other superior varieties.

(iv) Marker Assisted Breeding: With the progress in molecular biotechnology, various types of molecular markers has been developed during 1980s and 1990s in crop plants. The rapid development of molecular markers (particularly DNA markers) and continuous improvement of molecular assays has led to the birth of marker-assisted breeding (MAB).Molecular markers have greater opportunity for improving the efficiency of conventional breeding by carrying out selection not directly on the trait of interest but also on linked markers of that particular trait. Different types of markers are there but mostly used is the SSR which is not only reliable but also co-dominant in nature (Kumar *et al.* 2015).Compared with conventional breeding methods, MAB has significant advantages. (1) It allows selection for all kinds of traits at seedling stage and reduces the time required before the phenotype of an individual plant is known. (2) There is no effect of environment, thus allowing theselection to be performed under any environmental conditions. (3) Codominant markers allow effective selection of recessive alleles of desired traits in the heterozygous status. (4) For the traits controlled by multiple genes/QTLs, individualgenes/QTLs in the same individuals can be identified and selected simultaneously and thus suitable for gene pyramiding. (5) There is also higher effectiveness and efficiency in terms of time and resources. Genomic DNA fingerprinting is used as reliable tool for identification of bacterial blight (Onasanya *et al.*, 2010). By the use of DNA finger printing using molecular markers various isolates of *Xanthomonas oryzae* is easily identified which also helps in knowing the migration of these isolates from one place to another and thus helps in proper management (Onasanya *et al.*, 2011). With the help of molecular markers a total of 24 genes conferring resistance to bacterial blight has been identified. These single genes when used provide resistance to single isolate of pathogen but with the time new virulent race of pathogen appears which leads to the breakdown of resistance. Hence, convergent breeding technique came in which several genes are pyramided into single cultivar. Through MAB genes for bacterial blight resistance like *Xa 4, xa 5, xa 13* and *Xa 13* has been incorporated into multiple background with non-additive effect combination of *Xa 21 + xa 13 + xa 5, Xa 21 + xa 5, Xa 21+ xa 13* possess high level of resistance (Mishra *et al.*, 2016). It is more efficient, effective and reliable than phenotypic selection which can shorten the development time of varieties significantly, so in some cases it will be more cost effective than selection based on phenotypes.In case of rose resistant to black spot, three Rdr genes (Von Malek and Debener 1998; Whitaker *et al.*, 2010; Hattendorf *et al.*, 2004) has been identified. While in case of powdery mildew one dominant resistant gene, Rpp1, which provides complete resistance to race 9 of powdery mildew (Linde and Debener 2003; Linde *et al.*, 2004) and multiple QTLs in various genetic sources has been reported (Yan *et al* 2006; Moghaddam *et al.* 2010; Moghaddam *et al.*, 2012; Leus *et al.*, 2010. Xu *et al.*, 2005; Xu *et al.*, 2007; Debener and Byrne 2014, reported markers for two QTLs for powdery mildew resistance in *Rosa roxburghii.* One of the major limitations in using this method is high of cost and low reliability of markers to determine phenotype.

MAB has been used for identification of some of the diseases in wheat

Trait	Locus	Marker	Chromosome	Reference
Leaf rust	*Lr 21*	Ksu-D14	1DS	Talbert *et al.*, 1994
	Lr34	csLVLr34		Lagudah *et al.*, 2006
Stem rust	*SrCad*	*FSD-RSA*	6D	Hiebert *et al.*, 2011
	Sr39	*Sr39-F2*		Gold *et al.*, 1999
	Sr40	*Wmc344-F*		Wu *et al.*, 2009
	Lr24-Sr24	*12-F*		Mago *et al.*, 2005
Stripe rust	*Lr37-Yr17-Sr38*	*VENTRIUP/LN2*	2AS	Helguera *et al.*, 2003
Fusarium	Fhb: Qfhs.ndsu-3BS	Gwm 493	3BS	Randhawa *et al.*, 2013
head	Fhb2	Wmc 398	6B	Cuthbert *et al.*, 2007
blight	Fhb-5AS	Gwm 293	5AS	Mc Cartney *et al.*, 2004
Common bunt	Bt10	FSD-RSA	6D	Laroche *et al.*, 2000
	Blizzard	Gwm 374	1B	Wang *et al.*, 2009
		Gwm 264		
		Barc 128		
Loose smut	Utd1	Gwm 234	5BS	Randhawa *et al.* 2009
		Gwm 443		

(v) Transgenic strategies: It is an advanced form of genetic engineering in which foreign genes are introgressed into the genome of plant species for achieving desired goal. In this method transgenic plants acquire a single desired trait without any alteration of the original genetic makeup which allows multiple gene insertion simultaneously into genome to obtain broad-spectrum resistant lines. It a novel strategy through which fungal diseases can be effectively managed by introducing and overexpressing of genes that encode proteins which are involved in the synthesis of compounds toxic to fungi and with direct inhibitory effect on the growth of fungi (Cornelissen and Melchers 1993). Some strategies for improving blast resistance based on the host–pathogen gene-for-gene interaction system and antifungal protein genes have been developed (Campbell *et al*., 2002; Tan *et al*., 2004). There are also some of the reports on increasing rice blast resistance through transformation of chitinase gene (Nishizawa *et al*., 1999), plant antitoxin gene (Stark-Lorenzen *et al*., 1997), chitinase–glucanase gene (Feng *et al*., 1999). It is used as an attractive alternative to conventional techniques for the genetic improvement of Basmati rice (Jain and Jain 2000) in which resistance against blast diseases and improved nutritional quality (accumulation of pro-vitamin A and essential aminoacids in endosperm) has been done. In case of viral infestation like plum pox virus detection and its diagnosisis made possible (Lopez-Moya *et al*., 2000). Most successful and widely applied has been the exploitation of plant viral coat protein (CP) genes achieving CP-mediated resistance (Goldbach *et al.,* 2003). For a number of RNA viruses, including TMV, PVX, AlMV, CMV and TRV it has been shown that host plants transgenically expressing their CP results in high levels of operational resistance (Wilson, 1993; Beachy, 1994; Baulcombe, 1996), which can only be overcome at extremely high inoculation pressure or inoculation with unencapsidated viral RNA. In case of Prunus which is susceptible for plum pox virus is made resistant by transferring coat protein (Ravelonandro *et al*., 2000). Plants of *Prunus armeniaca* and *Prunus domestica* (Ravelonandro *et al*.,1997) transgenic lines have been reported to be resistant to PPV challenge, even when they were grafted over infected rootstocks.In case of stone fruit it is difficult to obtain virus (plum pox virus) resistant plant through classical plant breeding (Kegler *et al*., 1998). Hence with the time virus genome analysis has been developed (Lain *et al*.1988; Ravelonandro *et al*., 1988) and also improvement in woody plant species has been improved by transformation (Mante *et al*., 1991; Laimer da Camara Machado *et al*., 1992). Some of lines of *Nicotiana benthamiana* plants containing the P1 gene of PPV were resistant when challenged with the virus (Tavert-Roudet *et al*., 1998). Resistant lines responses to it in terms of amount of RNA detection suggesting a transgene silencing mechanism phenomenon occurred. In case of banana wilt caused by *Xanthomonas campestris* pv. *malvacearum* even if resistant germplasm is identified through conventional breeding it is difficult and lengthy process due to sterility of most of the cultivars coupled with long generation times. Hence, to overcome these problems transgenic technologies provide a cost effective alternative solutions to *Xanthomonas* wilt pandemic (Blomme *et al.*, 2014). It adds desirable trait against *Fusarium oxysporum* f. sp. *cubense*. (Smith *et al.*, 2014; Tripathi 2003; Remy *et al.*, 2013; Ortiz and Swennen, 2014). A new type of hypothetical system has been developed in tomato which includes two component systems and single component system. In two component system pathogen which produces non-specific elicitor bounds with the receptor produced by *cf* 9gene which results in hypersensitive response. In contrast to single component system when non-specific elicitor enters inside the cell interacts with RNAse and results in cell death (Singh, 2009).Alien gene transfer is an important means of developing new resistant germplasms and increasing the genetic diversity of wheat Fusarium head blight (FHB) resistance. Till today, only seven major FHB resistance QTLs have been reported (Li *et al.*, 2016): *Fhb1* (Cuthbert *et al.*, 2006; Liu *et al.*, 2006), *Fhb2* (Cuthbert *et al.*, 2007), *Fhb3* (Qi *et al.*, 2008), *Fhb4* (Xue *et al.*, 2010), *Fhb5*, *Fhb6* (Cainong *et al.* 2015), and *Fhb7* (Guo *et al.*, 2015). Of the seven permanently designated *Fhb* loci, three loci were transferred from wild relatives: *Fhb3* from *Leymus racemosus* (Qi *et al.*, 2008), *Fhb6* from *Elymus tsukushiensis* (Cainong *et al.*, 2015), and *Fhb7* from *Thinopyrum ponticum* (Guo *et al.*, 2015). Li *et al.*, 2016 reported *Thinopyrum intermedium* as a useful source of resistance to FHB, stripe rust and powdery mildew. The line L693 derived from the wild cross between common wheat and *Thinopyrum intermedium* (Liu *et al.*, 2015) exhibits strong FHB resistance.With the development of reverse genetics several genes have been identified which are associated with markers which has been used to transfer resistance to genetic backgrounds of susceptible variety which are high yielder (Brown *et al.*, 2016). It is also found that PR10 plays an important role in providing resistance to plants. In case of barley it was found *PR10* gene is activated upon infection by *Rhynchosporium secalis* (Steiner-Lange *et al.*, 2003). In cowpea, *PR10* homolog was found to be up regulated in resistant epidermal cells inoculated with the rust fungus *Uromyces vignae* (Mould *et al.*, 2003) and in case of rice *PR10* transcript was also induced in rice during infection by *Magnaporthe grisea* (McGee *et al.*, 2001). One of the technique found recently is RNA interference (RNAi) in which a cellular process found in most eukaryotes involved in developmental regulation of gene expression as well as defense against viruses and transposons (Fire *et al.*, 1998; Watson *et al.*, 2005; Mahmood-Ur-Rahman *et al.*, 2008; Jinek and Doudna, 2009; Obbard *et al.*, 2009; Hu *et al.*, 2015). RNAi is triggered by the presence of double stranded RNA (dsRNA) in the cell and ultimately leading to the degradation of homologous single stranded RNA molecules, inhibition of translation and modification of homologous genomic sequences (Qi and Hannon, 2005; Mahmood-Ur-Rahman *et al.*, 2008). Interference of double-stranded RNA with expression of specific genes has been widely described (Fire *et al.*, 1998). This post transcriptional gene silencing (PTGS) is a sequence-specific RNA degradation process triggered by a dsRNA, which propagates systemically throughout the plant, leading to the degradation of homologous RNA encoded by endogenous genes, and transgenes. ZmCORp, a protein identified against fungal conidia *A. flavus*, inhibited germination of conidia by 80% and decreased mycelial growth by 50%. The artificial manipulation of this pathway has allowed the efficient silencing of genes intransgenic plants and provided a powerful platform for reverse genetics studies (Duan *et al.* 2012). An alternative approach involving RNAi, known as Host-Delivered RNAi (HD-RNAi) or Host Induced Gene Silencing (HIGS), has emerged in recent years that use the host plant as a delivery system to cause gene silencing in the pathogen (Fairbairn *et al.*, 2007; Mahmood-Ur-Rahman *et al.*, 2008). In the HD-RNAi approach, the host plant is transformed with a hairpin construct targeting a pathogen gene. Once the transgenic plant is infected and the pathogen starts feeding from the host, the small interfering RNA (siRNA) molecules are transferred to the pathogen cells, ultimately inducing an RNAi response in the pathogen and the silencing of the targeted gene. Forward genetics studies have revealed a number of genes involved in *Fusarium* pathogenicity including argininosuccinate lyase (*ARG1*); class V and class VII chitin synthases (*CHSV* and *CHSVb*); two different heterotrimeric G protein α subunits (*FGA1* and *FGA2*); a heterotrimeric G protein β subunit (*FGB1*), a MAP kinase (*FMK1*) and a mitochondrial carrier protein (*FOW1*) (Di Pietro *et al.*, 2001; Namiki *et al.*, 2001; Inoue *et al.*, 2002; Jain *et al.*, 2002; Jain *et al.*, 2005; Madrid *et al.*, 2003; Martin-Urdiroz *et al.*, 2008).

(vi) Tissue culture Approach: Tissue culture plays an important role in selection of explants which are selected either through toxins or callus culture. These toxins include phytotoxic metabolites which are generated by pathogens which may be from pathogenic fungi or bacteria. These pathogenic fungi belongs from *Fusarium*, *Phytopthora*, *Alternaria* and *Helminthosporium* while pathogenic bacteria is comes under the genera *Clavibacter*, *Erwinia*, *Pseudomonas* and *Xanthomonas*. As reported by many researchers these phytotoxins are recognized as a useful tool for screening and breeding for disease resistant plants (Chawla and Wenzel, 1987; Vidyasekaran *et al.*, 1990). These toxins are classified into host specific and non-specific toxins. Among these metabolites fusaric acid and deoxynivalenol plays a significant role in disease development in plants. In case of culture filterate, plantlets are selected from survived explant in amended culture filterate and regenerated to get disease resistant plantlets. For example in case of banana wilt resistant plants are selected using culture filterate as a selection agent has been reported by Matsumoto *et al*., 1999. In grapevine these culture filterate found to have resistant and verified by Jaysankar *et al*., 2000; Jaysankar and Gray 2003. These toxin culture filtrates and purified toxins has been used for selection of disease resistant plants (Thakur *et al*. 2002; Li *et al*., 2005). *in vitro* toxin based selection approach has been described by Chandra *et al*., 2010 which is described in Fig. 1.

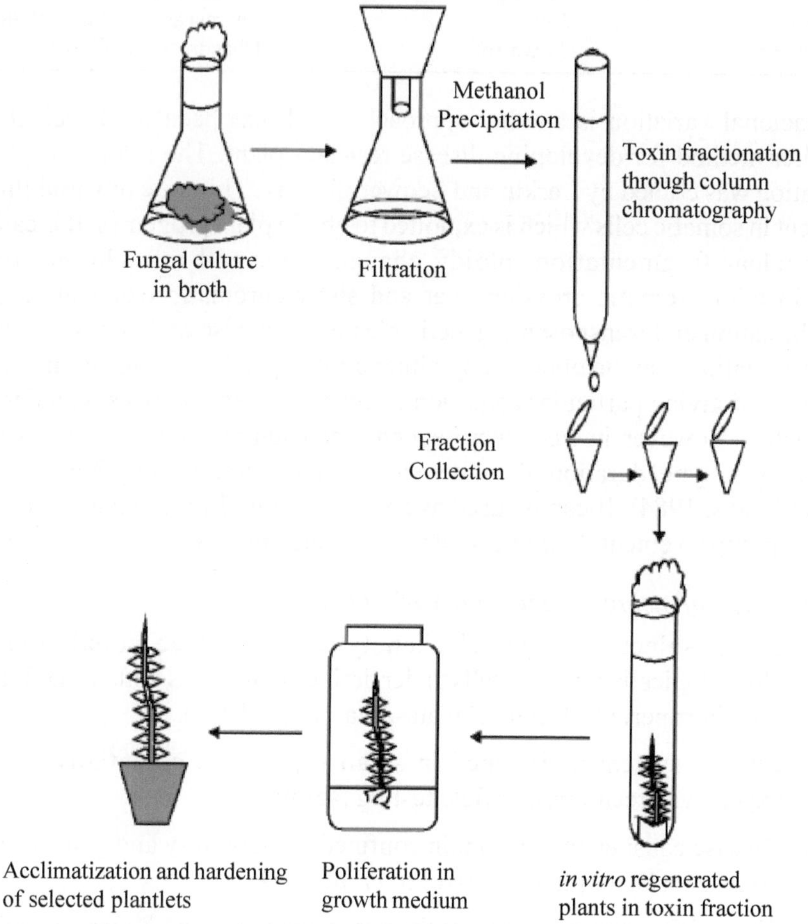

Fig. 1 : Steps to be followed for *in vitro* selection of disease resistant plants

It is considered more advantageous over culture filterate as specific toxins are used in selection process. This type of *in vitro* selection was first described by Behnke 1979 using toxins derived from culture filterate of pathogens. Fusaric acid isolated and purified from sp. race I was used for selecting multiple bud clumps to generate resistant line against wilt in banana (Widodo and Sudarsono, 2005).

Table 1: *In vitro* selection for disease resistance

Crops	Resistance	Reference
Apple	Black spot	Raman and Goodwin 2001
Banana	Wilt	Matsumoto *et al.* 1999
Guava	Wilt	Bajpai *et al.* 2007
Mango	Anthracnose	Jayasankar *et al.* 1999
Peach	Bacterial spot	Hammerschlag 1984
Pear	Fire blight	Duron *et al.* 1987
Papaya	Wilt	Sharma and Skidmore 1988
Strawberry	Crown rot	Chandra *et al.* 2010

Somaclonal variation is another approach which has been used recently as another method for developing disease resistant plant. The term somaclonal variation was coined by Larkin and Scowcroft 1981. This type of variability is present in somatic cells which is exploited to obtain plant progenies. It is caused by nuclear fragmentation, ploidy change, inter and intrachromosomal translocation, somatic crossing over and sister chromatid exchanges, gene amplification and transposable genetic elements. In case of *in vitro* condition these variations can be obtained by either applying selection agent on culture medium or giving particular condition to change the somaclones with desired characters. However, its frequency depends upon number of factors like explant source, genotype, duration of culture and medium composition (Brar and Jain 1998; Evans, 1984). It can be used as an excellent tool for inducing variation for crop improvement (Hammerschlag and Bauchan 1984; Vos *et al.* 1998).

Advantages of using somaclonal variations

(1) Direct selection of novel phenotypes from large populations of physiologically uniform cells under defined conditions with limited space and short period of time (Zapata-Arias *et al.* 1995).

(2) Culture system maintained in small spaces can potentially replace expensive greenhouse or field testing facilities.

(3) Disease causing agents remain confined to laboratory and generation of disease resistant plants (Brazolot *et al.* 1994).

(4) The cell free culture filtrate or the pure toxins of the pathogens and direct infection by the pathogen or all of them together could be used for the selection of novel disease resistant plants (El-Kazzaz and EL-Mougy, 2001; El-Kazzaz, 2001; El-Kazzaz and Ashour, 2004).

(5) Rapid development of resistant lines from existing commercial cultivars is easily fulfilled.

Table 2 : *in vitro* culture of somaclonal variants

Crops	*in vitro* culture	Pathogen	Reference
Apple	Callus culture	*Venturia inequalis*	Beech and Gessler 1986
	Mesophyll cells	*Phytopthora cactorum*	Mezzetti *et al.*, 1992
Banana	Adventitious bud	*Fusarium oxysporum* f. sp. *cubense* race 4	Hwang 1990
Grapevine	Protoplast	*Bortrytis cinerea*	Reustle and Matt 2000
	Somatic embryos	*Elsinoe ampelina*	Jayasankar *et al.*, 2000
Peach	Zygotic embryo calli	*Xanthomonas campestris*	Hammerschlag 1990
Mango	Embryogenic culture	*Colletotrichum gleosporioides*	Jaysankar *et al.*, 1999
Strawberry	Shoot apex calli	*Alternaria alternate*	Takahashi *et al.*, 1992

In case of sugarcane breeding tissue culture plays an important role as in case of conventional breeding there is slow multiplication rate, limited availability of seeds and perpetuation of diseases from generation to generation. Through tissue culture sugarcane varietyPhule savitri has been developed resistance to red rot and smut (Jalaja *et al.*, 2006), Ono has been developed which provides resistance against Fiji disease. Similarly, isolation of somaclones with Fiji and downy mildew resistance (Krishnamurthi and Tlaskal, 1974; Krishnamurthi, 1975; Tlaskal, 1975), smut resistance (Liu and Chen, 1978; Sreenivasan *et al.*, 1987a), rust resistance (Sreenivasan *et al.*, 1987b) and eye spot resistance (Heinz *et al.*, 1977; Larkin and Scowcroft, 1983) has been reported. Izquierdo *et al.* 1988 and Peros *et al.*, 1989 regenerated plantlets having high tomoderate resistance to rust (*Puccinia melanocephala*) disease using leaf explants of susceptible plants. Coleman 1981 initiated cultures from fifteen sugarcane varieties susceptible to mosaic virus disease and regenerated numerous somaclones, out of them 116 somaclones were found resistant to this disease. Tissue culture plays an important role in improving the plant quality and producing virus and phytoplasma free sugarcane plants from infected lines. Apart from that cryotherapy and micropropogation using the shoot apical meristem as the explants plays an important role in obtaining virus free plantlets (Tiwari *et al.*, 2010). Meristem culture was successfully used to eliminate ratoon stunting disease, sugarcane mosaic virus (SCMV), chlorotic streak disease, white leaf disease (Leu, 1978) and sugarcane yellow leaf virus (SCYLV) (Tiwari *et al.*, 2008). In combination with heat treatment, meristem and callus cultures results in effective production of pathogen-free stocks from plants infected with Fiji disease virus (FDV) (Wagih *et al.*, 1995), downy mildew (Leu, 1978), and SCMV. Recently it was reported that direct plant regeneration using thin cell layer culture helps in rapid production of disease-free plants from sugarcane infected with FDV and SCMV. Cryotherapy of shoot tips was found to be an efficient method for elimination of sweet potato little leaf (SPLL) phytoplasma from sweet potato. Cryotherapy along with thermotherapy helps in elimination of Raspberry bushy dwarf virus (RBDV). Cryotherapy also helps in long term storage of germplasm and production of virus and phytoplasma free plants (Wang and Valkonen, 2007). In case of tomato DNAP-17 has been developed showing resistance against Fusarium. In celery UC-T3 is having resistance against *Fusarium oxysporium* and in potato Russet Burbank having resistance against *Alternaria solani*, *Phytophthora infestans*. Similarly, clone no 7185 has been developed in potato which provide resistance against *Phytophthora infestans* (Rosenberg *et al.*, 2010). Ghag *et al.*, 2015 isolated somaclonal variants of banana (cv. *Rasthali*) from 14 years long continuously maintained embryogenic cultures derived from banana shoot tip which showed good resistance against *Fusarium oxysporium* var.*cubense* race 1.

Conclusion

Starting from the beginning of human civilization diseases covers a long journey from early man to the modern civilized man. With the advancement of life forms and with the change of environment several diseases have occurred in endemic to epidemic form. Some of the most famous examples are late blight of potato, powdery mildew of grapes, Bengal famine, coffee rust and recently a new rust has been identified in Uganda *i.e.* Ug99 in wheat. These diseases mainly occurred since previously focus was given on conventional approach which includes mass selection in which varieties are selected on the basis of their phenotypic performance. Apart from that with the time other methods like pedigree, backcross, recurrent selection, interspecific breeding came into existence. But with the advancement of new technologies and with the discovery of molecular markers it paved a new pathway for discovering the disease resistant traits linked to the molecular markers. Various QTLs has been identified in several crops through which it paved a new pathway for development of disease resistant varieties. An efficient cost-effective MAB technology has been developed that will allow breeders to assess the genotype across the full genome to recombine genes of agronomic importance from diverse sources (Jena and Mackill, 2008). Through gene pyramiding when a variety is developed it is resistant against multiple diseases. Not only, through pyramiding but also through transgenic strategies and with the discovery of forward and reverse breeding various genes has been silenced against several diseases. Apart from that tissue cultures also plays a significant role by release of variety resistant against several diseases through cell or toxin culture filtrate. Apart from cryotherapy and thermotherapy varieties have been also released through somaclonal variations which helps in elimination of several types of viral diseases. Thus through the present study we can say that we should not only focus to a particular method for release of disease resistant variety as with the time various advanced races of pathogens are evolving. So, we have to look every nook and corner of the breeding fields and have to study the mechanism of their virulence and then to think for a new goal which fulfil our desire of controlling them effectively.

References

Abenes M.L.P., Angeles E.R., Khush G.S., Huang N. (1993). Selection of bacterial blight resistant plants in the F2 generation via their linkage to molecular markers. *Rice Genet Newsl.* 10: 120–123.

Adugna A. (2004). Alternative approaches in deploying genes for disease resistance in crop plants. *Asian Journal of Plant Sciences.* 3(5): 618-623.

Ahloowalia B.S., Maluszynski M., Nichterlein K. (2004). Global impact of mutation-derived varieties. *Euphytica.* 135(2):187–204.

Aminu K.C. (1940). Interspecific hybridization between Asiatic and New world cottons. *Ind. J. Agric. Sci.* 10: 404-412.

Anjum Z.I., Hayat K., Chalkin S., Azhar T.M., Shehzad U., Ashraf F., Tariq M., Mehmood H.T., Azam M. (2015). Development of cotton leaf curls virus tolerance varieties through interspecific hybridization. *African Journal of Agricultural Research.* 10(13): 1612-1618.

Annonymous, http://www.iancas.org/oct07.pdf

Azlan S., Alias I., Saad A., Habibuddin H. (2004). Performance of potential mutant lines of MR 180. In: Sivaprasagam *et al.* (eds) Modern rice farming. Proceedings of the international rice conference Serdang, Malaysia: MARDI, 2003, pp 293–296

Bajpai A., Chandra R., Mishra M. and Tiwari R.K. (2007). Regenerating *psidium* spp. For screening wilt resistance rootstock under *in vitro* conditions. *Acta Hortic.*535: 145-154.

Baulcombe D.C. (1996). Mechanisms of pathogen-derived resistance to viruses in transgenic plants. *Plant Cell.* 8 (10): 1833-1844.

Beachy R.N. (1994). Introduction: transgenic resistance to plant viruses. *Semin. Virol.* 4: 327-328.

Beech, I. and Gessler C. (1986). Interaction between *Venturia inequalis* and apple callus tissue cultures: An electron microscopic study. *J. Phytopathology.*, 115: 315-322.

Behnke M. (1979). Selection of potato callus for resistance to culture filtrates of *Phytophthora infestans* and regeneration of resistant plants. *Theor. Applied Genet.*, 55: 69-71.

Blomme G., Jacobsen K., Ocimati W., Beed F., Ntamwira J., Sivirihauma C., Ssekiwoko F., Nakato V., Kubiriba J., Tripathi L., Tinzaara W., Mbolela F., Lutete L., Karamura E. (2014). Fine-tuning banana Xanthomonas wilt control options over the past decade in East and Central Africa. *Eur J Plant Pathol.* 139:271–287

Bonman J.M., Khush, G.S., Nelson, R.J. (1992). Breeding rice for resistance to pests. *Ann Rev Phytopathol.* 30:507–528.

Borlaug N.E. (1958). The use of multilineal or composite varieties to control air-borne epidemic disease of self pollinated crop plants. *In: Proc. First Intl. Wheat Genet. Symp.* University of Manitoba, Winnipeg, pp.: 12-31.

Brar DS and Jain SM (1998). Somaclonal Variation: Mechanism and Applications in Crop Improvement. In: Somaclonal Variation and Induced Mutations in Crop Improvement, Jain, S.M., D.S. Brar and B.S. Ahloowalia (Eds.). Kluwer Academic Publisher, London, pp: 15.

Brazolot J., Yu KF and Pauls KP (1994). *in vitro* Selection for Disease Resistance. In: Plant Cell Culture a Practical Approach, Dixon, R.A. and R.A. Gonzales (Eds.). 2nd Edn., IRL Press, Oxford, New York, Tokyo, pp: 87-98.

Brinkerhoff L.A. (1970). Variation in *Xanthomonas malvacearum* and its relation to control. *Ann. Rev. Phytopathol.* 8: 85-110.

Brown RL, Menkir A, Chen ZY, Luo M and Bhatnagar D (2016). Identification of gene Markers in aflatoxin-resistant maize germplasm for marker-assisted breeding. *Aflatoxins–Biochemistry and Molecular Biology*. 91-106.

Browning J.A., Frey K.J. (1969). Multiline cultivars as a means of disease control. *Ann. Rev. Phytopathol.*, 7: 355-382.

Burdon J.J. (1993). Genetic Variation in Pathogen Populations and Its Implications to Adaptation to Host Resistance. In: Jacobs, Th. and J.E. Parlevliet (Eds.). Durability of Disease Resistance. Kluwer Academic Publishers, Dordrecht, pp. 41-56.

Cainong, J. C., Bockus, W. W., Feng, Y., Chen, P., Qi, L., Sehgal, S. K., Danilova, T. V., Koo, D.-H., Friebe, B. and Gill, B. S. (2015). Chromosome engineering, mapping, and transferring of resistance to Fusarium head blight disease from Elymus tsukushiensis into wheat. *Theor. Appl. Genet.* 128:1019-1027.

Campbell M.A., Heather A.F., Pamela C.R. (2002). Engineering pathogen resistance in crop plants. *Transgenic Res.* 11(6):599–613.

Chandra R, Kamle M, Bajpai A, Muthukumar M and Kalim S. (2010). *In vitro* selection: a candidate approach for disease resistance breeding in fruit crops. *Asian Journal of Plant Sciences*., 9(8): 437-446.

Chawla HS and Wenzel G (1987). *in vitro* selection of barley and wheat for resistance against *Helminthosporium sativum. Theory and Applied Genetics*, 74: 841-845.

Coleman, R. R. (1981). New plants produced from callus tissue culture. In: Sugarcane Res. 1981 Report, US Dept. *Agric. Res. Ser.* p. 38.

Cornelissen B.J.C., Melchers L.S. (1993). Strategies for control of fungi diseases with transgenic plants. *Plant Physiol.* 101:709–712.

Courtois B., Nelson R., Roumen E. (1997). Creation of a gene pool to improve Piricularia on Resistance partial secanano rice through recurrent selection. In: Guimaraes EP (ed) Recurrent selection in rice. International Center for Tropical Agriculture, Cali, pp 189–202.

Cuthbert, P. A., Somers, D.J. and Brulé-Babel, A. (2007). Mapping of Fhb2 on chromosome 6BS: a gene controlling Fusarium head blight field resistance in bread wheat (*Triticum aestivum* L.). *Theor. Appl. Genet.* 114:429-437.

Cuthbert, P. A., Somers, D. J., Thomas, J., Cloutier, S., and Brulé-Babel, A. (2006). Fine mapping Fhb1, a major gene controlling Fusarium head blight resistance in bread wheat (*Triticum aestivum* L.). *Theor. Appl. Genet.* 112:1465-1472.

Da Camara Machado A., Knapp E., Puhringer H, Seifert G, Hanzer V, Weiss H, Wang Q, Debener T and Byrne DH (2014). Review: Disease resistance breeding in rose: Current status and potential of biotechnological tools. *Plant Science*, 8968: 1-11.

Di Pietro, A., Garcia-Maceira, F. I., Meglecz, E., and Roncero, M. I. G. (2001). A MAP kinase of the vascular wilt fungus *Fusarium oxysporum* is essential for root penetration and pathogenesis. *Mol. Microbiol*, 39: 1140–1152.

Duan, C.-G.,Wang, C.-H., and Guo, H.-S. (2012). Application of RNA silencing to plant disease resistance. *Silence*, 3:5.

Duron N., Paulin JP and Brisset MN. (1987). Use of *in vitro* propagation plant material for rating fire blight susceptibility. *Acta Hortic*., 217: 317-324.

El-Kazzaz A.A. (2001). Inheritance of resistance to *Fusarium oxysporum* f.sp. *lycopersici*in F2 tomato plants via tissue culture. *Egypt J Genet Cytol*. 30:51–59

El-Kazzaz A.A., Ashour A.M.A. (2004). Genetically resistant cucumber plants to wilt pathogen via tissue cultures. *Egypt J. Phytopathol*. 32:1–10

El-Kazzaz A.A., EL-Mougy N.S. (2001). Inheritance of disease resistance in cucumber plants to root rot caused by Fusarium solani using tissue culture techniques. *Egypt J Phytopathol*. 29:57–68.

Ellis MH, Stiller WN, Phongkham T, Tate WA, Gillespie VJ, Gapare WJ, Zhu QH, Llewellyn DJ and Wilson IW. (2016). Molecular mapping of bunchy top disease resistance in *Gossypium hirsutum* L. *Euphytica*. 210(1): 135-142.

Evans D.A., Sharp W.R., Medina-Filho H.P. (1984). Somaclonal and gametoclonal variation. *Am J Bot*. 71:759–774.http://www.jstor.org/stable/2443467

Fairbairn, D. J., Cavallaro, A. S., Bernard, M., Mahalinga-Iyer, J., Graham, M. W., and Botella, J. R. (2007). Host-delivered RNAi: an effective strategy to silence genes in plant parasitic nematodes. *Planta*. 226, 1525–1533.

Feng D.R., Wei J.W., Xu X.P., Xu Y., Li B.J. (1999). Introduction of multiple antifungal protein genes into rice and preliminary study on resistance to *Pyricularia oryzae* of transgenic rices. *Acta Sci Nat Univ Sunyatsen*. 38(4):62–66

Fire, A., Xu, S. Q., Montgomery, M. K., Kostas, S. A., Driver, S.E., and Mello, C.C. (1998). Potent and specific genetic interference by double-stranded RNA in *Caenorhabditis elegans*. *Nature* 391: 806–811.

Ghag S.B., Shekhawat U.K.S., Ganapathi T.R. (2015). Fusarium wilt of banana: biology, epidemiology and management. *International Journal of Pest Management*, 61 (3):250-263.

Gnanamanickam S.S., Priyadarisini V.B., Narayanan N.N., Vasudevan P., Kavitha, S. (1999). An overview of bacterial blight disease of rice and strategies for its management. *Current Science*. 77:1435–1444.

Gold J, Harder D, Townley-Smith F, Aung T, and Procunier J. (1999). Development of a molecular marker for rust resistance genes *Sr39* and *Lr35* in wheat breeding lines. *Electron. J. Biotechnol*. 2: 1-6.

Goldbach R., Bucher E, Prins M. (2003). Resistance mechanisms to plant viruses: an overview. *Virus Research*. 92: 207-212.

Gould, F. (1983). Genetics of Plant Herbivore Systems: Interactions between Applied and Basic Study. In: Denno, R.F. and M.S. McClure (Eds.). Variable Plants and Herbivores in Natural and Managed Systems. *Academic Press*, New York, pp.: 599-653.

Guimaraes E.P., Correa-Victoria F. (2000). Use of recurrent selection for develop resistance Pyricularria grisea Sacc. On rice. In: Guimaraes EP (ed) Advances in rice population improvement. Embrapa Rice and Beans, Santo Antonio de Goias, pp. 165–175.

Guo, J., Zhang, X., Hou, Y., Cai, J., Shen, X., Zhou, T., Xu, H., Ohm, H. W., Wang, H., Li, A., Han, F., Wang, H., and Kong, L. (2015). High-density mapping of the major FHB resistance gene Fhb7 derived from Thinopyrum ponticum and its pyramiding with Fhb1 by marker-assisted selection. *Theor. Appl. Genet*. 128:2301-2316.

Hammerschlag F.A. and Bauchan G.R. (1984). Genetic stability of callus derived from peach embryos. *Hotic. Sci*., 19: 78-78.

Hammerschlag, F.A. (1990). Resistant responses of plants regenerated from peach callus to *Xanthomonas campestris* pv. *pruni*. *J. Am. Soc*., 115: 1034-1037.

Hattendorf A, Linde M, Mattiesch L, Kaufmann H, Debener T. (2004). Genetic analysis of rose resistance genes and their localization in the rose genome. *Acta Hort*., 651: 123-130.

Heinz, D. J., Krishnamurthi, M., Nickell, L. G. and Maretzki, A. (1977). Cell, tissue and organ culture in sugarcane improvement, In: J. Reinert andY. P. S. Bajaj (eds): Applied and Fundamental Aspects of Plant Cell, Tissue and Organ Culture. Springer, Berlin, pp. 3–17.

Helguera M, Khan IA, Kolmer J, Lijavetzky D, Zhong-qi L, and Dubcovsky J. (2003). PCR assays for the *Lr37-Yr17-Sr38* cluster of rust resistance genes and their use to develop isogenic hard red spring wheat lines. *Crop Sci*. 43: 1839-1847.

Hiebert CW, Fetch TG, Zegeye T, Thomas JB, Somers DJ, Humphreys DG, McCallum BD, Cloutier S, Singh D and Knott D.R. (2011). Genetics and mapping of seedling resistance to Ug99 stem rust in Canadian wheat cultivars Peace and AC Cadillac. *Theor. Appl. Genet*. 122: 143-149.

Hu Z, Parekh U, Maruta N, Trusov Y and Botella J.R. (2015). Down-regulation of *Fusarium oxysporum* endogenous genes by Host-Delivered RNA interference enhances disease resistance. *Frontiers in Chemistry*.3, 1.

Hurni, S., Scheuermann, D., Krattinger, S. G., Kessel, B., Wicker, T., Herren, G., Fitze, M., Breen, J., Presterl, T., Ouzunova, M. and Keller, B. (2015). The maize disease resistance gene *Htn1* against northern corn leaf blight encodes a wall-associated receptor-like kinase. *Proc Natl Acad Sci*. U S A. 112(28):8780-8785.

Hwang, S.C., (1990). Somaclonal Resistance in Cavendish Banana to Fusarium Wilts. In: Fusarium Wilt of Banana, Ploetz, R.C. (Eds.). APS Press, St. Paul, pp: 121-125.

Inoue, I., Namiki, F., and Tsuge, T. (2002). Plant colonization by the vascular wilt fungus *Fusarium oxysporum* requires FOW1, a gene encoding a mitochondrial protein. *Plant Cell* 14: 1869–1883.

Izquierdo, F., Bravo, N. and Cutierrez, R. (1988). Early evaluation of rust resistance in sugarcane tissue derived plantlets. *Revistade Protection Vegetat*. 3: 217–226.

Jain R.K. and Jain S (2000). Transgenic strategies for genetic improvement of Basmati rice. *Indian Journal of Experimental Biology*. 38(1): 6-17.

Jain, S., Akiyama, K., Kan, T., Ohguchi, T., and Takata, R. (2003). The G protein beta subunit FGB1 regulates development and pathogenicity in *Fusarium oxysporym. Curr. Genet.* 43: 79–86.

Jain, S., Akiyama, K., Mae, K., Ohguchi, T., and Takata, R. (2002). Targeted disruption of a G protein alpha subunit gene results in reduced pathogenicity in *Fusarium oxysporum. Curr. Genet.* 41:407–413.

Jain, S., Akiyama, K., Takata, R., and Ohguchi, T. (2005). Signaling via the G protein alpha subunit FGA2 is necessary for pathogenesis in *Fusariumoxysporum*. *FEMS Microbiol. Lett.* 243:165–172.

Jalaja N.C., Sreenivasan T.V., Pawar S.M., Bhof P.G., Garker R.M. (2006). Co 94012- A new sugarcane variety through somaclonal variation. *Sugar Tech*. 8:132-136

Jayasankar S., Litz R.E. and Gray D.J. (2000). *In vitro* selection of *Vitis vinifera* chardonnay with *Elsinoe ampelina* culture filtrates is accompanied by fungal resistance and enhanced secretion of chitinase. *Planta*, 211: 200-208.

Jayasankar S., Litz R.E., Gray D.J. and Moon P.A. (1999). Responses of embryogenic mango cultures and seedling bioassays to a partially purified phytotoxin produced by a mango leaf isolate of *Colletotrichum gloeosporioides* Penz. *in vitro Cell. Devel. Biol. Plants*. 35: 475-479.

Jayasankar, S., Litz RE and Gray D.J. (2000). *in vitro* selection of Vitis vinifera Chardonnay with Elsinoe ampelina culture filtrates is accompanied by fungal resistance and enhanced secretion of chitinase. *Planta*. 211: 200-208.

Jena K.K., Mackill D.J. (2008). Molecular Markers and Their Use in Marker-Assisted Selection in Rice. *Crop Sci*., 48:1266–1276.

Jensen N.E. (1952). Intravarietal Diversification in Oat Breeding. *Agron. J*., 44: 30-34.

Jinek, M., and Doudna, J. A. (2009). A three-dimensional view of the molecular machinery of RNA interference. *Nature*, 457: 405–412.

Jorgensen J.H. (1993). Durability of Resistance in the Pathosystem: Barley Powdery Mildew. In: Jacobs, Th. and J.E. Parlevliet (Eds). Durability of Disease Resistance. *Kluwer Academic Publishers*, Dordrecht, pp: 159-176.

Joseph M., Gopalakrishnan S., Sharma R.K., Singh V.P., Singh A.K., Singh N.K., Mohapatra T. (2004). Combining bacterial blight resistance and Basmati quality characteristics by phenotypic and molecular marker-assisted selection in rice. *Mol Breed*. 13(4): 377–387.

Kegler H., Fuchs E., Gruntzig, M., Schwarz S. (1998). Some results of 50 years of research on the resistance to plum pox virus. *Acta Virologica*. 42: 200–215.

Khush G.S. (2005). What it will take to Feed 5.0 Billion Rice consumers in 2030. *Plant Molecular Biology*. 59:1–6.

Knight R.L.(1957). Black arm disease of cotton and its control. In the "Plant protection conference 1956". Butterworths Scientific Publications, London, pp. 53-59.

Krattinger, S. G., Sucher, J., Selter, L. L., Chauhan, H., Zhou, B., Tang, M., Upadhyaya, N. M., Mieulet, D.,Guiderdoni, E., Weidenbach, D., Schaffrath, U., Lagudah, E. S.and Keller, B. (2016). The wheat durable, multipathogen resistance gene *Lr34* confers partial blast resistance in rice. *Plant Biotechnol J.* 14(5):1261-1268.

Krishnamurthi M and Tlaskal J. (1974). Fiji disease resistant *Saccharum officinarum* var. Pindar sub-clones from tissue cultures. *Proc. Int. Soc. Sugar Cane Technol.* 15:130–137.

Krishnamurthi, M. (1975). Note on disease resistance of tissue culture subclone and fusion of sugarcane protoplast. *Sugarcane Breed. Newsl.* 35, 24–26.

Kumar, R., Mandal, S.S., Mishra, A.K., Smriti, Singh, R., Kumar, P. (2015). Heterosis and combining ability for yield and its contributing traits of kharif maize (*Zea mays* L.). *The Bioscan*. 10 (4): 2049-2056

Lagudah ES, McFadden H, Singh RP, Huerta-Espino J, Bariana HS and Spielmeyer W. (2006). Molecular genetic characterization of the Lr34/Yr18 slow rusting resistance gene region in wheat. *Theor. Appl. Genet.* 114: 21-30.

Laimer da Camara Machado, M., da Camara Machado, A., Hanzer, V., Weiss, H., Regner, F., Steinkellner, H., Mattanovich, D., Plail, R., Knapp, E., Kalthoff, B., Kattinger, H. (1992). Regeneration of transgenic plants of *Prunus armeniaca* containing the coat protein of Plum Pox Virus. *Plant Cell Rep.* 11: 25–29.

Lain S., Riechmann J.L., Mendez E., Garcia J.A. (1988). Nucleotide sequence of the 3% terminal region of plum pox potyvirus RNA. *Virus Res.* 10: 325–342.

Larkin P.J., Scowcroft W.R. (1981). Somaclonal variation-a novel source of variability from cell cultures for plant improvement. *Theor Appl Genet.* 60(4):197–214.

Larkin, P. J. and Scowcroft,W. R. (1983). Somaclonal variation and eyespot toxin tolerance in sugarcane. *Plant Cell Tissue Organ Culture*, 2: 111–121.

Laroche A, Demeke T, Gaudet DA, Puchalski B, Frick M and McKenzie R (2000). Development of a PCR marker for rapid identification of the Bt-10 gene for common bunt resistance in wheat. *Genome*, 43: 217-223.

Leu, L. S. (1978). Apical meristem culture and redifferentiation of callus masses to free some sugarcane systemic diseases. *Plant Protect. Bull.* (Taiwan) 20: 77–82.

Leus L, Moghaddam HH and Van Huylenbroeck J. (2010). Pathotype specific powdery mildew resistance in roses. *Acta Hort.* 870: 99-102.

Li S., Wang H., Zhang J., Fitt B.D., Xu Z., Evans N., Liu Y., Yang W., Guo X. (2005). *in vitro* mutation and selection of doubled-haploid Brassica napus lines with improved resistance to *Sclerotinia sclerotiorum. Plant Cell Rep.* 24: 133-144.

Li X, Xiang ZP, Chen WQ, Huang QL, Liu TG, Li1 Q, Zhong SF, Zhang M, Guo JW, Lei L, Luo P.G. (2016). Reevaluation of Two QTLs for Type II Resistance to *Fusarium* HeadBlight in Wheat Germplasm PI 672538. *Phytopathology*. DOI: 10.1094/PHYTO-04-16-0170-R.

Lillemo, M., Joshi, A.K., Prasad, R., Chand, R. and Singh, R. P. (2013). QTL for spot blotch resistance in bread wheat line Saar co-locate to the biotrophic disease resistance loci *Lr34* and *Lr46*. *Theor Appl Genet.* 126:711-719.

Linde M, Mattiesch L and Debener T. (2004). Rpp1, a dominant gene providing race specific resistance to rose powdery mildew (Podosphaera pannosa): molecular mapping, SCAR development and confirmation of disease resistance data, *Theor. Appl. Genet.* 109: 1261-1266.

Linde M. and Debener T. (2003). Isolation and identification of eight races of powdery mildew of roses (Podosphaera pannosa) (Wallr.: Fr.) de Bary and the genetic analysis of the resistance gene Rpp1, *Theor. Appl. Genet.* 107: 256-262.

Liu, M. C. and Chen, W. (1978). Tissue and cell culture as aids to sugarcane breeding-II. Performance and field potential of Callus derived lines. *Euphytica*, 27: 273–282.

Liu, S., Zhang, X., Pumphrey, M.O., Stack, R.W., Gill, B.S. and Anderson, J. A. (2006). Complex microcolinearity among wheat, rice, and barley revealed by fine mapping of the genomic region harboring a major QTL for resistance to Fusarium head blight in wheat. *Functional & Integrative Genomics*. 6(2):83-89.

Lopez-Moya J.J., Fernandez-Fernandez M.A., Cambra M., Garcia J.A. (2000). Biotechnological aspects of plum pox virus. *Journal of Biotechnology*, 76: 121–136

Madrid, M. P., Di Pietro, A., and Roncero, M. I. G. (2003). Class V chitin synthase determines pathogenesis in the vascular wilt fungus *Fusarium oxysporum* and mediates resistance to plant defence compounds. *Mol. Microbiol.* 47: 257–266.

Mago R, Bariana HS, Dundas IS, Spielmeyer W, Lawrence GJ, Pryor AJ, and. Ellis JG (2005). Development of PCR markers for the selection of wheat stem rust resistance genes *Sr24* and *Sr26* in diverse wheat germplasm. *Theor. Appl. Genet.* 111: 496-504.

Mahmood-Ur-Rahman, Ali, I., Husnain, T., and Riazuddin, S. (2008). RNA interference: the story of gene silencing in plants and humans. *Biotech. Adv.* 26: 202–209.

Mante S., Morgens P.H., Scorza R., Cordts J.M., Callahan A.M. (1991). Agrobacterium-mediated transformation of plum (Prunus domestica L.) hypocotyl slices and regeneration of transgenic plants. *Bio: Technology*. 9: 752–758.

Marshal, D.R. (1977). The advantages and hazards of genetic homogeneity. Ann. N.Y. *Acad. Sci.*, 287: 1-20.

Martin-Urdiroz, M., Roncero, M.I.G., Gonzalez-Reyes, J.A. and Ruiz-Roldan, C. (2008). ChsVb, a class VII chitin synthase involved in septation, is critical for pathogenicity in *Fusarium oxysporum*. *Eukaryot. Cell.*, 7: 112–121.

Matsumoto K, Barbosa ML, Souza LAC and Teixeira J.B. (1999). *In vitro* selection for *Fusarium* wilt resistance to banana II. Resistance to culture filterate of race 1 *Fusarium oxysporum* f sp. *cubense*. *Fruits*, 54: 151-157.

McCartney CA, Somers DJ, Fedak G, and Cao W. (2004). Haplotype diversity at fusarium head blight resistance QTLs in wheat. *Theor. Appl. Genet.* 109: 261-271.

McGee, J.D., Hamer, J.E. & Hodges, T.K. (2001). Characterization of a PR-10 pathogenesis related gene family induced in rice during infection with *Magnaporthe grisea*. *Molecular and Plant Microbe Interactions,* Vol. *14*, pp. 877–886

Mezzetti B., Zimmerman R.H., Mischke C., Rosati P. and Hammerschlag F.A. (1992). Merocyanine 540 as an optical probe to monitor the effects of culture filtrates of *Phytophthora cactorum* on apple cell membrane. *Plant Sci.*, 83: 163-167.

Miah G., Rafii M.Y., Ismail M.R., Puteh A.B., Rahim H.A., Asfaliza R. and Latif M.A. (2013). Blast resistance in rice: A review of conventional breeding to molecular approaches. *Mol. Biol. Rep*. 40: 2369–2388.

Mishra, A.K., Singh, P.K., Kumar, A., Kumar, M., Kumar, R., Singh R., Singh, S.P., (2016). Bacterial leaf blight resistance through marker assisted approach among elite rice (*Oryza sativa* L.) varieties. International Symposium on Management of rice based agricultural System under stress prone environment.

Moghaddam HH, Leus L, de Riek J, Van Huylenbroeck J and Van Bockstaele E. (2012). Construction of a genetic map with SSR, AFLP and morphological markers to locate QTLs controlling pathotype-specific powdery mildew resistance in diploid roses. *Euphytica*, 184: 413-427.

Moghaddam HH, Leus L, Van Huylenbroeck J, Van Bockstaele E and de Riek J. (2010). Pathotype dependent resistance mapping for powdery mildew in a diploid rose population. *Acta Hort.* 870: 103–107.

Mohamad O., Nazir B.M., Alias I., Azlan S., Abdul Rahim H., Abdullah M.Z., Othman O., Hadzim K., Saad A., Habibuddin H., Golam F., (2006). Development of improved rice varieties through the use of induced mutations in Malaysia. *Plant Mutat Rep.* 1(1):27–33.

Mohanraj D, Padmanaban P and Karunakaran M. (2003). Effect of phytotoxin of *Colletotrichum falcatum* went. (*Physalophora tucumanensis*) on sugarcane in tissue culture. *Acta Phytopathol. Entomol. Hung*., 38: 21-28.

Mould, M.J., Xu, T., Barbara, M., Iscove, N.N. & Heath. M.C. (2003). cDNAs generated from individual epidermal cells reveal that differential gene expression predicting subsequent resistance or susceptibility to rust fungal infection occurs prior to the fungus entering the cell lumen. *Molecular and. Plant Microbe Interactions,* 16: 835–845

Namiki, F., Matsunaga, M., Okuda, M., Inoue, I., Nishi, K., Fujita, Y., (2001). Mutation of an arginine biosynthesis gene causes reduced pathogenicity in *Fusarium oxysporum* f. sp *melonis.Mol. PlantMicrobe Interact,* 14: 580–584.

Nishizawa Y., Nishio Z., Nakazono K., Soma M., Nakajima E., Ugaki M., Hibi T. (1999). Enhanced resistance to blast (*Magnaporthe grisea*) in transgenic Japonica rice by constitutive expression of rice chitinase. *Theor Appl Genet*. 99(3-4):383–390.

Obbard,D. J., Gordon, K.H. J., Buck, A.H., and Jiggins, F.M. (2009). The evolution of RNAi as a defence against viruses and transposable elements. *Philos. Trans. R. Soc. Lond. B Biol. Sci.* 364: 99–115.

Onasanya A., Basso A., Somado E., Gasore E.R., Nwilene F. E., Ingelbrecht I., Lamo J., Wydra K., Ekperigin M.M., Langa M., Oyelakin O., Sere Y., Winter S., Onasanya R.O. (2010). Development of a combined molecular diagnostic and DNA fingerprinting technique for rice bacterial pathogens in Africa. *Biotechnology*, 9: 89-105.

Onasanya A., Kiepe P., Basso A., Nkima G., Nwilene F.E., Ingelbrecht I., Lamo J., Ekperigin M.M., Onasanya R.O., Oyelakin O., Winter S., Sere Y.(2011). Molecular characterization and DNA fingerprinting of *Xanthomonas oryzae* pv. *Oryzae* isolates from climatic change prone areas in East Africa. *Research journal of environmental sciences*. 5(9): 772-778.

Ortiz R., Swennen R. (2014). From crossbreeding to biotechnology- facilitated improvement of banana and plantain. *Biotechnol Adv*. 32:158-169.

Paterson A.H., Damon S., Hewitt J.D., Zamir D., Rabmowitch H.D., Lincoln S.E., Lander E.S., Tanksley S.D. (1991). Mendelian factors underlying quantitative traits in tomato: comparison across species, generations and environments. *Genetics*, 127(1): 181–197.

Peros, J. P., Bonnel, E. F. and Maubussian, J. C. (1989). Altered rust resistance and yield components arising from leaf tissue and bud culture in sugarcane. *Proc. 20th Congr. ISSCT,* pp. 732–738.

Qi, L. L., Pumphrey, M. O., Friebe, B., Chen, P. D., and Gill, B. S. (2008). Molecular cytogenetic characterization of alien introgressions with gene Fhb3 for resistance to Fusarium head blight disease of wheat. *Theor. Appl. Genet*. 117:1155-1166.

Qi, Y. J., and Hannon, G. J. (2005). Uncovering RNAi mechanisms in plants: biochemistry enters the foray. *FEBS Lett.* 579: 5899–5903.

Raman, H. and Goodwin P.B. (2000). *In vitro* screening of apple germplasm for resistance against black spot caused by *Venturia inaequails*. *J. New Seeds*, 2: 37-48.

Randhawa HS, Asif MD, Pozniak C, Clarke JM, Graf RJ, Fox SL, Humphreys DG, Knox RE, De Pauw RM, Singh AK, Cuthbert RD, Hucl P and Spaner D. (2013). Review: Application of molecular markers to wheat breeding in Canada. *Plant Breeding*, 132: 45-471.

Randhawa HS, Popovic Z, Menzies J, Knox R and Fox S (2009). Genetics and identification of molecular markers linked to resistance to loose smut (*Ustilago tritici*) race T33 in durum wheat. *Euphytica*, 169: 151-157.

Ravelonandro M., Scorza R., Bachelier J.C., Labonne G., Levy L., Damsteegt V., Callahan A.M., Dunez J. (1997). Resistance of transgenic Prunus domestica to plum pox virus infection. *Plant Dis*. 81:1231–1235.

Ravelonandro M., Scorza R., Callahan A., Levy L., Jacquet C., Monsion M., Damsteegt V., (2000). The use of transgenic fruit trees as a resistance strategy for virus epidemics: the plum pox (sharka) model. *Virus Research,* 71: 63–69.

Ravelonandro, M., Dunez, J., Scorza, R, Labonne, G. (1998). Challenging transgenic plums expressing potyvirus coat protein genes with viruliferous aphids. *Acta Horticulturae.* 472: 413–420.

Remy S., Kovacs G., Swennen R., Panis B. (2013). Genetically modified bananas: past present and future. *Acta Horticulturae*. 974:71-80.

Reustle, G.M. and Matt A. (2000). First step to use the protoplast technique for breeding purposes. *Acta Horticulturae*, 528: 341-346.

Risk, J.M., Selter, L.L., Chauhan, H., Krattinger, S.G., Kumlehn, J., Hensel, G., Viccars, L.A. (2013). The wheat Lr34 gene provides resistance against multiple fungal pathogens in barley. *Plant Biotechnol. J.* 11: 847–854.

Risk, J.M., Selter, L.L., Krattinger, S.G., Viccars, L.A., Richardson, T.M., Buesing, G., Herren, G. (2012). Functional variability of the Lr34 durable resistance gene in transgenic wheat. *Plant Biotechnol. J.* 10, 477–487.

Rosenberg V., Tsahkna A., Kotkas K., Tähtjärv T., Särekanno M., Liiv K. (2010). Somaclonal variation in potato meristem culture and possibility to use this phenomenon in seed potato production and breeding. Agronomy Research. 8 (Special Issue III): 697–704.

Sharma, N.K. and Skidmore D.I. (1988). *in vitro* expression of partial resistance to *Phytophthora palmivora* by shoot cultures of papaya. *Plant Cell Tissue Organ Culture*, 14: 187-196.

Shu Q., Wu D., Xia Y. (1997). The most widely cultivated rice variety 'Zhefu 802' in China and its geneology. *MBNL*. 43:3–5

Singh P. (2013). Essentials of Plant Breeding. Kalyani Publication., pp. 209

Singh, B.D.(2009). Plant Breeding: Principles and Method. Kalyani Publication, pp. 486.

Smith M.K., Langdona P.W., Pegg K.G., Daniells J.W. (2014). Growth, yield and Fusarium wilt resistance of six FHIA tetraploid bananas (*Musa* spp.) grown in the Australian subtropics. *Scientia Horticulturae*. 170:176-181.

Sreenivasan, J., Sreenivasan, T. V. andAlexander, K. C. (1987a). Somaclonal variation for rust resistance in sugarcane. *Indian. J. Genet*. 47: 109–114.

Sreenivasan, J., Sreenivasan, T.V., Alexander, K. C. and Rao, M. (1987b). Somaclonal variation for smut disease (Ustilago scitaminea Syd.) resistance in sugarcane. In: G. M. Reddy (ed.): *Proc. Symp. Plant Cell Tissue Culture Economically Important Plants*. Hyderabad, pp. 259–264.

Sreewongchai T., Toojinda T., Thanintorn N., Kosawang C., Vanavichit A., Tharreau D., Sirithunya P. (2010). Development of elite indica rice lines with wide spectrum of resistance to Thai blast isolates by pyramiding multiple resistance QTLs. *Plant Breed.* 129:176–180.

Stark-Lorenzen P., Nelke B., HanBler G., Muhlbach H.P., Thomzik J.E. (1997). Transfer of a grapevine stilbene synthase gene to rice (*Oryza sativa* L). *Plant Cell Rep.* 16(10):668–673.

Steiner-Lange, S., Fischer, A., Boettcher, A., Rouhara, I., Liedgens, H., Schmelzer, E. & Knogge, W. (2003). Differential defense reactions in leaf tissues of barley in response to infection by *Rhynchosporium secalis* and to treatment with a fungal avirulence gene product. *Molecular and Plant Microbe Interactions,* Vol. 16, pp. 893– 902.

Sucher J, Boni R, Yang P, Rogowsky P, Büchner H, Kastner C, Kumlehn J, Krattinger SG and Keller B. (2016). The durable wheat disease resistance gene *Lr34* confers common rust and northern corn leaf blight resistance in maize. *Plant Biotechnology*. DOI: 10.1111/ pbi.12647.

Takahashi H., Takatsugu T. and Tsutomu M. (1992). Resistant plants to *Altemaria alternata* strawberry pathotype selected from calli clones of strawberry cultivar Morioka-16 and their characteristics. *J. Jap. Soc. Hort. Sci.*, 61: 323-329.

Talbert LE, Blake NK, Chee PW, Blake TK and Magyar GM. (1994). Evaluation of sequence-tagged-site pcr products as molecular markers in wheat. *Theor. Appl. Genet*. 87: 789-794.

Tan Y.N., Yi Z.L., Jiang J.X., Qin J.P., Xiao L. (2004). Strategies and advances in improving resistance to rice blast by transgenic approaches. *Mol. Plant Breeding*. 2(6):847–852.

Tavert-Roudet G., Ravelonandro M., Bachelier J.C., Dunez, J., 1998. Transgenic Nicotiana benthamiana plants containing the P1 gene of plum pox virus are resistant to virus challenge. *Eur. J. Plant Pathol*. 104, 103–107.

Thakur, Sharma MD and Sharma S. (2002). *In vitro* selection and regeneration of carnation (*Dianthus caryophyllus* L.) plants resistant to culture filtrate of *Fusarium oxysporum* f. sp. *dianthi*. *Plant Cell Rep*., 20: 825-828.

Tiwari AK, Bharti YP, Tripathi S, Mishra N, Lal M, Rao GP, SharmaPK and Sharma ML (2010). Review Article: Biotechnological Approaches to Improve Sugarcane Crop with Special Reference to Disease Resistance. *Acta Phytopathologica et Entomologica Hungarica*, 45(2): 235–249.

Tiwari, A. K.,Yadav, S., Tripathi, S., Rastogi, J., Bharti,Y. P., Lal, M. and Sharma, M. L. (2008). Performance of micro propagated, SCYLV infected and healthy plants of sugarcane. *Indian J. Sugarcane Technology*, 23: 41–44.

Tlaskal, J. (1975). A note on disease resistant subclones derived from tissue culture techniques. *Sugarcane Breed, Newsletter*, 36: 71–72.

Toojinda T., Tragoonrung S., Vanavichit A., Siangliw J.L., Pa-In N., Jantaboon J., Siangliw M., Fukai S. (2005). Molecular breeding for rainfed lowland rice in the Mekong region. *Plant Prod Sci*., 8(3): 330–333.

Tripathi L. (2003). Genetic engineering for improvement of *Musa* production in Africa. *Afr J Biotechnol*, 2:503-508.

Vidyasekaran P, Ling DH, Borromeo ES, Zapata FJ and Mew TW. (1990). Selection of brown resistant rice plants from *Helminthosporium oryzae* toxin resistant calluses. *Annals of Applied Biology*,117: 515-523.

Von Malek B and Debener T. (1998). Genetic analysis of resistance to blackspot (Diplocarpon rosae) in tetraploid roses, *Theor. Appl. Genet*., 96: 228-231.

Vos JE, Schoeman MH, Berjak P, Waft MP and Toerien A.J. (1998). *In vitro* selection and commercial release of guava wilt resistant rootstocks. *Acta Hortic*, 513: 69-79.

Wagih, M. E., Gordon, G. H., Ryan, C. C. and Adkins, S. W. (1995). Development of an axillary bud culture technique for Fiji disease virus elimination in sugarcane. *Aust. J. Bot*. 43: 135–143.

Wang S, Knox RE, DePauw RM, Clarke FR, Clarke JM and Thomas J.B. (2009). Markers to a common bunt resistance gene derived from Blizzard wheat (*Triticum aestivum* L.) and mapped to chromosome arm 1BS. *Theor. Appl. Genet*., 119: 541-553.

Wang, Q. and Valkonen, J. P. T. (2007). Elimination of virus and phytoplasma by cryotherapy of *in vitro* grown shoot tips. 1st Meeting on "Cryopreservation of Crop Series in Europe", pp. 58–59.

Watson, J.M., Fusaro, A.F., Wang, M. B. and Waterhouse, P. M. (2005). RNA silencing platforms in plants. *FEBS Lett*. 579: 5982–5987.

Welz, H.G. and Geiger, H. H. (2000). Genes for resistance to northern corn leaf blight in diverse maize populations. *Plant Breeding*. 119, 1-4.

Whitaker VM, Debener T, Roberts AV, Hokanson S.C. (2010). A standard set of host differentials and unified nomenclature for an international collection of Diplocarpon rosae races. *Plant Pathology*, 59: 745-752.

Widodo Y. and Sudarsono. (2005). *in vivo* selection of peanut somatic embryos on medium containing culture filtrates of *Sclerotium rolfsii* and plantlet regeneration. *Hayati*, 12: 50-56.

Wilson T.M.A. (1993). Strategies to protect crop plants against viruses: pathogen-derived resistance blossoms. *Proc. Natl. Acad. Sci.* USA 90: 3134-3141.

Wolfe M.S., Barrett J.A., Slater, S.E. (1983). Pathogen Fitness in Cereal Mildews. In: Lamberti, F., J.M. Waller, N.A. Van der Graff (Eds.). Durable Resistance in Crops. Plenum Press, New York and London, pp. 27-37.

Wolfe M.S. (1985). The current status and prospects of multiline cultivars and variety mixtures for disease resistance. *Ann. Rev. Phytopathology.*, 23: 251-273.

Wu S, Pumphrey M and. Bai G. (2009). Molecular Mapping of Stem- Rust Resistance Gene *Sr40* in Wheat. *Crop Sci.* 49: 1681-1686.

Xu Q, Wen X and Deng X. (2007). Cloning of two classes of PR genes and the development of SNAP markers for powdery mildew resistance loci in chestnut rose (*Rosa roxburghii* Tratt). *Molecular Breeding.* 19: 179-191.

Xu Q, Wen XP and Deng X.X. (2005). Isolation of TIR and non TIR NBS-LRR resistance gene analogues and identification of molecular markers linked to a powdery mildew resistance locus in chestnut rose (*Rosa roxburghii* Tratt). *Theor. Appl. Genet.* 111: 819-830.

Xue, S., Li, G., Jia, H., Xu, F., Lin, F., Tang, M., Wang, Y., An, X., Xu, H., Zhang, L., Kong, Z., and Ma, Z. (2010). Fine mapping Fhb4, a major QTL conditioning resistance to Fusarium infection in bread wheat (*Triticum aestivum* L.). *Theor. Appl. Genet.* 121:147-156.

Yan Z, Dolstra O, Prins TW, Stam P and Visser P.B. (2006). Assessment of partial resistance to powdery mildew (Podosphaera pannosa) in a tetraploid rose population using a spore-suspension inoculation method. *European Journal of Plant Pathology.* 114: 301–308.

Zapata-Arias F.J., Torrizo L.B., Ando A. (1995). Current developments in plant biotechnology for genetic improvement: the case of rice (*Oryza sativa* L.). *World J. Microbiol Biotechnol.*, 11(4): 393-399.

2

Insights of the Mechanism of Male Sterility in Plants

Rishav Kumar[1] and Sachin Tyagi[2]

[1]Department of Plant Breeding and Genetics
Bihar Agricultural University, Sabour, Bhagalpur-813210, Bihar India
[2]Department of Horticulture (Fruit and Fruit Technology)
Bihar Agricultural University, Sabour, Bhagalpur-813210, Bihar, India

Introduction

With the change in global scenario there is increase in pressure of meeting the food requirements of rapid growing population but due to scarcity of resources it is becoming very difficult to meet the needs of present population but even if there is so much problem still nature has provided solution to sort out from it through hybrid seed production through male sterility. As male sterility is a condition in which pollen grain is either absent or non-functional in the pollen grain. Its first case was reported by Koelreuter in 1763. Later many scientist have reported about its role in hybrid seed production. Mainly the male sterility is characterized by malformation of androecium, lack of normal anther sac, inability of pollen to mature or to release from the anther sac and inability to develop normal microspore or pollen. Due to such malformation male sterility is characterized into phenotypic male sterility, genotypic male sterility and chemically induced male sterility. Later in 1998, Kaul further generalised the phenotypic male sterility into structural, sporogenous and functional male sterility. In case of structural male sterility stamens are completely absent or if present then it is either malformed or modified into other floral part. While in case of sporogenous male sterility a case of abnormal microsporogenesis has been reported. In functional male sterility pollen development is normal but its function is impaired.Genotypic male sterility which is further classified into genetic, cytoplasmic and cytoplasmic genetic male sterility.

Genetic male sterility: Sterility is governed by nuclear gene action which is mainly recessive (Lewis and Crow 1956) and governed by single gene but a case of two or more gene control has been also reported in several crop plants like wheat, maize, cotton, tomato, sugarbeet. But in case of safflower it is governed by dominant gene. GMS consists of two lines A line (male sterile line *mm*), and B line (maintainer line *Mm*). Here male sterility is governed by crossing sterile line with fertile line which yields 50% sterile and 50% fertile line. But it has certain constraints which limits its use in hybrid seed production i.e. it is less stable and since 50% progeny are male fertile so it increases the cost of hybrid seed production.GMS is further classified into Temperature sensitive genetic male sterility in which rice line PA64S shows sterility when temperature is above 27 °C and fertility when temperature is less than 23.5 °C. Similarly, rice line *Ugp 1* shows sterility when temperature is above 28 °C and fertility when it is below 21 °C. In case of Photoperiod sensitive genetic male sterility rice line shows sterility when it receives photoperiod more than 13.75 hrs and fertility when it is less than 13 hrs. In case of wheat line Norin 26 when photoperiod is more than 15 hrs then it is sterile while fertility when it is less than 14.5 hrs. A case of reverse PGMS has been also reported in rice line CSA which shows sterility when it receives photoperiod less than 12.5 hrs and fertility more than 13.5 hrs. Both TGMS and PGMS is governed by two recessive genes. Generally, TGMS is preferred over PGMS since when there is large variation in latitude then only it can be used while TGMS is used in all the countries which are accessible for cultivation.

In transgenic genetic male sterility a gene is introduced into the genome of an organism through recombinant DNA technology which is responsible for inducing male sterility. Its successful result has been reported in rapeseed and tobacco. In this system two types of transgenes are used barnase (A line) which causes male sterility and barstar (R line) which restores fertility. Both the genes has been obtained from *Bacillus amyloliquefaciens.* The male sterile line is identified with the help of another gene *bar* gene. The *barnase* gene is linked with the *bar* gene obtained from *Streptomyces.* This *bar* gene provides resistance to the herbicide phosphinohricin. This *barnase-bar/-* gene is used for construction of sterility gene while *barstar* gene is used for construction of fertility restorer gene. Here male sterile lines are maintained by crossing it with male fertile line resulting 1 male sterile : 1 male fertile and later fertile lines are eliminated by spraying phosphinothricin at seedling stage. It has been successfully used in maize, cauliflower, tomato, wheat and chicory.

Expression in *Brassica oleracea*: *Bcp1* gene found in *Brassica campestris* is expressed both in tapetum and microspore which results in normal pollen development. This gene is linked to pollen specific promoter *LAT52* which was obtained from *Arabidopsis thaliana.* An antisense construct of *Bcp1* was prepared and linked with hormone inducible enhancer sequence which was transferred into *Brassica oleracea.* This transgenic line expressed sterility only when it is sprayed with concerned hormone otherwise, in absence of its spray it showed sterility.

Expression in maize: Reported by Wu *et al.*, 2016 pollen grains with seed production technology transgene exhibited starch depletion due to expression of á-amylase and thus unable to germinate which acts as a male sterile while pollen grains which do not carry the seed production technology were non-transgenic and are able to fertilize homozygous mutant plants, resulting innon-transgenic male sterile progeny which acts as female parents.

Cytoplasmic male sterility: Sterility is controlled by the cytoplasmic gene or plasmagene which results due to mutation in the mitochondrial genome resulting into mitochondrial dysfunctioning. Mainly mutations plays an important role in changing the open reading frames creating chimeric genes by duplications and rearrangement of mitochondrial DNA sequence which results in the polypeptide formation in mitochondria affecting the major function resulting male sterility. Changes in the chloroplast genome has been reported in tobacco, barley, sorghum and rapeseed. CMS is maintained by crossing the cytoplasmic male sterile (A line) with the male fertile (B line) resulting male sterile lines. On regular backcrossing till 6-7 generations resulting the nuclear genotype of male sterile line would be completely identical to the recurrent parent hence, it plays an important role in producing hybrid seed in ornamental species and in those species where vegetative part is important.

Cytoplasmic-genetic male sterility: Sterility is governed by both cytoplasmic and nuclear genes. It was first discovered by Jones and Davis in 1944 in onion. Here three lines are required i.e. A (male sterile), B (maintainer) and R (restorer) line. It is maintained by crossing A line with B line producing 1 male sterile : 1 male fertile line. Male sterile is again crossed with restorer i.e. R line. But some problems occurs while utilizing it in hybrid seed production like

i. **Undesirable effect of cytoplasm:** With the development of hybrids new problem arising day by day in the form of undesirable effect of cytoplasm. Although, cytoplasm effects maize hybrids in the form of earlier silking, delayed pollen shedding, increases yield by 2-4% but it also makes mitochondria of maize highly susceptible to the T toxin produced by *Helminthosporium maydis* causing leaf blight in maize. Due to such undesirable effect it is not so much successful in other crops like tobacco.

ii. **Unsatisfactory fertility restoration:** Some cases have been also reported were fertility restoration is not satisfactory.

iii. **Unsatisfactory pollination:** This is problematic case which arises in hybrid crops when pollinators are required for carrying out pollination.

iv. **Spontaneous reversion:** In pearl millet it has been reported that there is very low frequency of spontaneous reversion so that in each generation of the CMS line pollen shedding plants are present which is very problematic in hybrid seed production.

v. **Modifying genes:** Some of the genes when transferred through backcross programme disturbs the nuclear genetic background leading to pollen production in male sterile plants.

vi. **Contribution of cytoplasm by sperm:** In tobacco it has been reported that when transgene is integrated into the chloroplast genome it showed 0.5% level of pollen transmission.

vii. **Environmental effects:** In maize, bajra and jowar there is partial breakdown of sterility under certain environmental conditions resulting in pollen production by male sterile plants.

viii. **Non-availability of suitable restorer line:** In some crops it has been reported that restorers are not easily available due to which it effects hybrid seed production.

But there are certain approaches which has been adopted to minimize the undesirable effect,

a. **Alternative source of CMS source:** Generally, hybrids are developed from single CMS source due to which it becomes after sometime as in southern corn leaf blight of maize but it is being minimized when used CMS-C and CMS-S instead of CMS-T. In case of sunflower it has been reported that when PET-1 cytoplasm is used then it doesn't have any adverse effect on yield of plant.

b. **Correction of the defect of male sterile cytoplasm:** When cytoplasmic hybrids are formed in *Brassica napus* and *Brassica juncea* through transfer of CMS-*ogu*then it shows chlorotic symptoms at seedling stage at low temperature. The defect of *ogura* cytoplasm were governed by chloroplast genome while male sterility is governed by mitochondrial genome. Such defects are corrected by somatic hybridisation.

c. **Research for new and superior nuclear gene combinations that alleviate the defects of the cytoplasm:** Generally, male sterile lines are effected by the defect in the nuclear genetic background of male sterile plants as some cases of spontaneous reversion has been reported in some crops. In such cases such maintainer should be used which eliminate the effect.

d. **Production of cytoplasmic blends hybrid:** Such hybrids are to be produced which consists of two types of seeds each having distinct types of cytoplasm i.e. in first case hybrid seeds are to be produced using CMS that shows partial restoration in fertility and in second case hybrid seeds are to be produced by detasseling or by manual emasculation.

e. **Use of alternative method of emasculation:** When CMS sources are not available then alternative method of emasculation is to be used. As in case of maize when CMS-T was used continuously then it leads to epidemic hence it was stopped growing for next 3 years and CMS- C and CMS- S was grown on small scale. Later, with the development of CHA several hybrids were developed.

Chemically Induced Male Sterility

Here chemicals are used to induce artificial non-genetic male sterility which effects the function of male reproductive organs in the plants. Its first case was reported in maize by Moore and Naylor in 1950 who used maleic hydrazide to induce male sterility. Later Laibach and Kribban used α-NAA and β-IAA to increase the proportion of staminate of male flowers in cucumber.There are certain properties due to which any CHA is considered ideal for its application on plants i.e. highly selective, should not be mutagenic, reproducible, economical, produces complete sterility, minimum reduction in seed set and no hazardous to environment.

Mode of action of CHA

a) Disruption of meiosis leading to the arrest of anther development and degeneration of pollen mother cells.

b) Disruption of exine formation resulting in thin walled, irregular and non-viable microspores.

c) Decrease in starch deposition and appearance of abnormal vacuoles in the microspores making them non-viable.

d) Abnormal growth of tapetal layers.

e) Indehiscence or delayed dehiscence of normally developed anthers with viable pollen.

f) Non-germination of pollen on stigma or cessation of elongation of pollen tube resulting in non-fertilization.

g) Affects DNA synthesis before the entry of cells into meiosis. Gametocytes acts on the chromosome before the entry of cells in prophase I which ultimately delays condensation of chromosome resulting delayed bivalent formation.

Stages of application of CHA: Mainly the damage is effective when applied at proper stage.

a) **Panicle initiation stage:** Affects normal development of anthers and pollen mother cells. If plasmodesmata is disturbed which connects sporogenous and tapetal cells it results in failure of development of pollen mother cells. Sometimes over growth of tapetum prior to callus formation blocks development of pollen mother cells ultimately resulting sterility in plants.

b) **Bolting and flowering stage:** Disturbs normal development and functioning of microspores. Tapetal cells border the development of pollen mother cells which provides nutrition for the development of pollen grains and any disturbance to its development results in male sterility.

Generative nucleus which is formed after division of microspore nucleus lies near the intine wall but when CHA is applied then it prevents it to lie near intine wall or to divide and thus makes pollen sterile (Colhoun and Steer, 1982). Some of CHA used and the their critical stages for application are presented in the table 1.

Table 1: Some of CHA used and the their critical stage for application

CHA	Critical stage for application	Crop
MG1, MG2	5 days before heading	Rice
DPX3778	Late boot stage	Wheat, triticale
Shell	Booting stage	Wheat, oat, barley, maize
Maleic hydrazide	Early microsporogenesis	Maize, wheat, cotton, onion
LY195259	Foliar spray when spike of main shoot is 5mm	Wheat
Mendok	Panicles wrapped in the flag leaf on the main tiller	Pearl millet

But there are certain shortcomings in CHA i.e. pollen abortion is incomplete, effective only at a specific stages of crop growth, ovule fertility is also adversely affected, interfere with cell division, deformation of leaf, stunting of plant growth, residual effect and high cost of gametocide.

Various changes occurred during male sterility

1. **Cytological changes:** Differentiation of sporogenous tissue, cell division continues without formation of archesporium, accumulation of undifferentiated parenchymatous cells in anther locule, lack of wall layers in defective anthers indicate failure of sporogenous cells to differentiate. Even when sporogenous tissue escape from the action of mutant gene the phenotype has limited internal differentiation. CHA effects the carbohydrate metabolism and thus starves tapetum which ultimately leads to degeneration of microspore.Changes like vacuolation, synthesis of callosic cell wall, formation of tubular extension on mitochondria, disorganisation of cytoplasm, disintegration of plastids and mitochondria, formation of darkly stained lipid, abortion of microsporocytes. Fusion between microsporocytes leads to formation of multinucleate plasmodial sporocytes, asynchronous and delayed meiosis, disturbance in orientation and pairing of homologous chromosome, failure of chiasmata formation, coagulation of pachytene threads and their dissolution, agglutination of bivalents and appearance of nuclei as structure less spherical bodies, desynapsis at diplotene, rupture of nuclear membrane at arrested pachytene stage, cytoplasmic disintegration of microsporocytes and disturbance of exine differentiation are observed.
2. **Biochemical Changes:** The level of proline, leucine, isoleucine, phenylalanine, valine is reduced while there is increase in asparagine, glycine, alanine, arginine, aspartic acids. Male sterile anthers contain lower protein content and fewer polypeptide bands. Polyamines like putrescine, spermidine and spermine is increased.

Enzymes: Callase activity is disturbed resulting in premature or delayed release of meiocytes and microspore. Activity of esterase is decreased which retards pollen development. Amylases function is decreased which corresponds with high starch content and reduced levels of soluble sugars. Accumulation of adenine due to the decrease in adenine phosphoribosyl transferase activity which is toxic for the development of microspores.

Models for the Mechanism of Male Sterility

A. **Cytotoxicity Model:** CMS proteins causes mitochondrial dysfunctioning in the sporophytic or gametophytic cells of the anthers leading to male abortion. In this model CMS protein directly kills the cells.

B. **Energy Deficiency Model:** CMS proteins results in mitochondrial deficiencies and cell fails to meet substantial energy requirement of male reproductive development. These protein may integrate into the inner mitochondrial membrane destroys the proton gradient affecting ATP synthesis.

C. **Aberrant Programme Cell Death Model:** Plant PCD functions in development of different structures. Release of cytochrome c from mitochondria into the cytosol and overproduction of Reactive Oxygen Species triggers apoptosis and plant PCD which causes degeneration of tapetum.

D. **Retrograde Regulation Model**: Sporophytic cells or the gametes dies due to abnormality in the redox processes and energy production in the mitochondria which causes retrograde signalling for Programme Cell Death in tapetum.

Role of mitochondria and nucleus in cytoplasmic male sterility: Here in case this case through retrograde signalling mitochondria through subunits of mitochondrial electron transfer chain (mtETC) complex interacts with other complexes by realising signals through reactive oxygen species in response to which cytochrome *c* activates abnormal PCD in tapetal cells or microspore which ultimately results in sporophytic or gametophytic sterility. Figure 1 shows the nuclear –mitochondria interaction and the role of RF genes in restoring the fertility in plants.

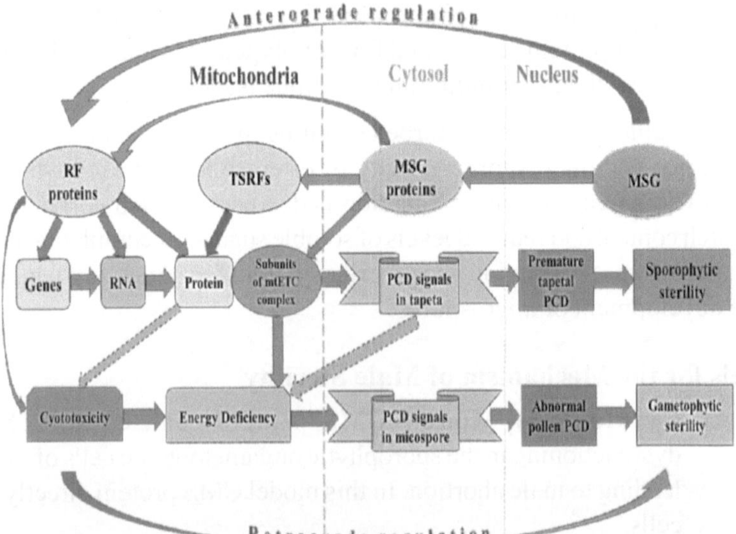

Fig. 1 : Nuclear-mitochondria interaction and the role of RF genes in restoring the fertility in plants

Role of restorer genes in restoring fertility: Through anterograde regulation nucleus generates mitochondrial sorting genes (MSG) which goes into cytosol to produce MSG proteins which ultimately produces tissue-specific regulatory factors (TSRFs) and RF proteins. TSRFs interacts with protein while RF proteins interacts at gene, RNA and protein level and supresses their expression thus restores fertility. The examples of some crops where nuclear-mitochondrial interaction leads to male sterility and CMS and their restorer are enlisted in the Table 5 and Table 6 respectively.

Table 2: Examples of some crops where nuclear-mitochondrial interaction leads to male sterility

Tobacco	Reduced rate of mitochondrial replication which limits ATP production leading to sterility. Occurrence of plastid enriched with amyloplast but with poorly developed grana in parenchyamtous cells of anther.
Sunflower	Breakdown of mitochondria results in fragmentation of endoplasmic reticulum and lack of dictyosome in tapetum.
Petunia	Presence of an array of vacuoles containing membranous inclusion and osmophilic material in tapetum.
Sugarbeet	Appearance of endoplasmic reticulum in concentric configurations in tapetum.

Table 3: CMS and their restorer

Crop	CMS type	ORF	Rf	References
Rice	CMS- BT	B-*atp6-orf79*	*Rf1a, Rf1b*	Akagi *et al.*, 2004, Kazama and Toriyama 2003, Wang *et al.*, 2006
	CMS-HL	*atp6-orfH79*	*Rf5*	Wang *et al.*, 2012, Wang *et al.*, 2013
	CMS-LD	L-*atp6-orf79*	*Rf2*	Itabashi *et al.*, 2009, Itabashi *et al.*, 2011
	CMS-CW	*orf307*	*Rf17*	Fujii *et al.*, 2007, Fujii *et al.*, 2010
	CMS-WA	*rpl5-WA352*	*Rf3,Rf4*	Zhang *et al.*, 1997
Maize	CMS-T	*urf13-atp4*	*Rf1, Rf2*	Dewey *et al.*, 1987, Dill *et al.*, 1997
	CMS-S	*orf355-orf77*	*Rf3*	Zabala *et al.*, 1997
	CMS-C	*atp6-C*	*Rf4*	Dewey *et al.*, 1991
Sunflower	CMS-PET1	*atp1-orf522*	*Rf1*	Kohler *et al.*, 1991
Brassica napus	CMS-Ogu	*orf138-atp8*	*Rfo*	Brown *et al.*, 2003
	CMS-Pol	*orf224-atp6*	*Rfp*	Mackenzie and McIntosh 1991, Singh and Brown 1991
	CMS-Nap	*orf222-nad5c-orf139*	*Rfu*	Brown 1999, L' Homme *et al.*,1997
Radish	CMS-Kos	*orf125-atp8*	*Rfk1*	Iwabuchi *et al.*, 1999, Koizuka *et al.*, 2003
Common bean	CMS-Sprite	*atp1-orf98-orf239*	*Fr2*	Abad *et al.*, 1995, He *et al.*, 1996, Mackenzie and McIntosh 1999
Sugarbeet	CMS-Owen	*preSatp6*	*Rf1*	Matsuhira *et al.*, 2012, Yamamoto *et al.*, 2005

References

Abad AR, Mehrtens BJ, Mackenzie SA. (1995). Specific expression in reproductive tissues and fate of amitochondrial sterility-associated protein in cytoplasmic male-sterile bean. *Plant Cell,* 7: 271–85.

Akagi H, Nakamura A, Yokozeki-Misono Y, Inagaki A, Takahashi H, Mori K, Fujimura T. (2004). Positional cloning ofthe rice *Rf-1* gene, a restorer of BT-type cytoplasmicmale sterility that encodes a mitochondria-targetingPPR protein. *Theor. Appl. Genet.* 108:1449–57.

Brown GG. (1999). Unique aspects of cytoplasmic male sterility and fertility restoration in *Brassica napus.J. Hered.*, 90:351–56.

Brown GG, Formanova N, Jin H, Wargachuk R, Dendy C, Patil P, Laforest M, Zhang J, Cheung WY, Landry BS. (2003). The radish *Rfo* restorer geneof Ogura cytoplasmic male sterility encodes a protein with multiple pentatricopeptide repeats. *Plant J.* 35:262–72.

Colhoum CW, Steer MW. (1982). Rev. *Cytol. Biol.Veg.*, 5:282-301.

Dewey RE, Timothy DH, Levings CS III. (1987). A mitochondrial protein associated with cytoplasmicmale sterility in the T cytoplasm of maize. *Proc. Natl. Acad. Sci.* USA 84: 5374–78.

Dewey RE, Timothy DH, Levings CS III. (1991). Chimeric mitochondrial genes expressed in the Cmale-sterile cytoplasm of maize. *Curr. Genet.* 20:475–82.

Dill CL, Wise RP, Schnable PS. (1997). Rf8 and Rf " mediate unique T-urf13-transcript accumulation, revealing a conserved motif associated with RNA processing and restoration of pollen fertility in Tcytoplasmmaize. *Genetics,* 147:1367–79.

Fujii S,KazamaT, YamadaM,ToriyamaK. (2010). Discovery of global genomic re-organization based oncomparison of two newly sequenced rice mitochondrial genomes with cytoplasmic male sterility-relatedgenes. *BMC Genomics* 11:209.

Fujii S, Komatsu S, Toriyama K. (2007). Retrograde regulation of nuclear gene expression in CW-CMSof rice. *Plant Mol. Biol.*, 63:405–17.

He S, Abad AR, Gelvin SB, Mackenzie SA. (1996). A cytoplasmic male sterility-associated mitochondrialprotein causes pollen disruption in transgenic tobacco. *Proc. Natl. Acad. Sci.* USA 93:11763–68.

Horn R, Kusterer B, Lazarescu E, PrufeM, Friedt W. (2003). Molecular mapping of the Rf1 gene restoringpollen fertility inPET1-basedF1 hybrids in sunflower (*Helianthus annuus* L.). *Theor. Appl.Genet.*, 106:599–606.

Itabashi E, IwataN, Fujii S, KazamaT,ToriyamaK. (2011). The fertility restorer gene, *Rf2*, for Lead Ricetypecytoplasmic male sterility of rice encodes a mitochondrial glycine-rich protein. *Plant J.*, 65:359–67.

Itabashi E, KazamaT,ToriyamaK. (2009). Characterization of cytoplasmicmale sterility of rice with Lead Rice cytoplasm in comparison with that with Chinsurah Boro II cytoplasm. *Plant Cell Rep.* 28:233–39.

Iwabuchi M, Koizuka N, Fujimoto H, Sakai T, Imamura J. (1999). Identification and expression of thekosena radish (*Raphanus sativus* cv. *kosena*) homologue of the ogura radish CMS-associated gene, *orf138. Plant Mol. Biol.*, 39:183–88.

Kazama T, Toriyama K. (2003). A pentatricopeptide repeat-containing gene that promotes the processingof aberrant *atp6* RNA of cytoplasmic male-sterile rice. *FEBS Lett.* 544:99–102.

Kohler RH, Horn R, Lossl A, Zetsche K. (1991). Cytoplasmic male sterility in sunflower is correlatedwith the co-transcription of a new open reading frame with the *atpA* gene. *Mol. Gen. Genet.*, 227:369–76.

Koizuka N, Imai R, Fujimoto H, Hayakawa T, KimuraY, Kohno-Murase J, Sakai T, Kawasaki S, Imamura J. (2003). Genetic characterization ofa pentatricopeptide repeat protein gene, *orf687,* that restores fertility in the cytoplasmic male-sterileKosena radish. *Plant J.* 34: 407-415.

L'Homme Y, Stahl RJ, Li X, Hameed A, Brown GG. (1997). *Brassica nap* cytoplasmic male sterility is associated with expression of a mtDNA region containing a chimeric gene similar to the *pol* CMS associated *orf224* gene. *Curr. Genet.* 31:325–35.

Mackenzie SA, McIntosh L. (1999). Higher plant mitochondria. *Plant Cell.,* 11:571–85.

Matsuhira H, Kagami H, KurataM, Kitazaki K,Matsunaga M, Hamaguchi Y, Hagihara E, Ueda M, Harada M, Muramatsu A, Yui-Kurino R, Taguchi K, Tamagake H, Mikami T, Kubo T. 2012. Unusual and typical featuresof a novel restorer-of-fertility gene of sugar beet (*Beta vulgaris* L.). *Genetics* 192:1347–58.

SinghM, Brown GG. (1991). Suppression of cytoplasmic male sterility by nuclear genes alters expressionof a novel mitochondrial gene region. *Plant Cell* 3:1349–62.

Uyttewaal M, Arnal N, Quadrado M, Martin-Canadell A, Vrielynck N, Hiard S, Gherbi H, Bendahmane A, Budar F, Mireau H. (2008). Characterization of *Raphanus sativus* pentatricopeptide repeat proteins encoded by the fertility restorer locus for Oguracytoplasmic male sterility. *Plant Cell* 20:3331–45.

Wang K, Gao F, Ji Y, Liu Y, Dan Z, Yang P, Zhu Y, Li S. (2013). ORFH79 impairs mitochondrial function via interactionwith a subunit of electron transport chain complex III in Honglian cytoplasmic male sterile rice. *New Phytol.* 198:408–18.

Wang Z, Zou Y, Li X, Zhang Q, Chen L, Wu H, Su D, Chen Y, Guo J, Luo D, Long Y, Zhong Y, Liu YG. (2006). Cytoplasmic male sterility of rice with boroII cytoplasm is caused by a cytotoxic peptide and is restored by two related PPR motif genes via distinct modes of mRNA silencing. *Plant Cell* 18:676–87.

Wu Y, Fox TW, Trimnell MR, Wang L, Xu R, Cigan AM, Huffman GA, Garnaat CW, Hershey H, Albertsen MC. (2016). Development of a novel recessive genetic male sterility system for hybrid seed production in maize and other cross-pollinating crops. *Plant Biotechnology Journal* 14: 1046–1054.

Yamamoto MP, Kubo T, Mikami T. (2005). The 5'-leader sequence of sugar beet mitochondrial atp6encodes a novel polypeptide that is characteristic of owen cytoplasmicmale sterility. *Mol. Genet. Genomics* 273:342–49.

Zabala G, Gabay-Laughnan S, Laughnan JR. (1997). The nuclear gene Rf3 affects the expression of themitochondrial chimeric sequence R implicated in S-type male sterility in maize. *Genetics,* 147:847–60.

Zhang, G, Lu, Y, Bharaj, TS, Virmani SS, Huang, N. (1997). Mapping of the *Rf-3* nuclear fertility-restoringgene for WA cytoplasmic male sterility in rice using RAPD and RFLP markers. *Theor. Appl. Genet.* 94:27–33.

3

Biotechnological Approaches for Crop Improvement

M Hamida Nasrin[1] and Israr Ahmad[2]

[1]*Central Island Agricultural Research Institute, Port Blair, Andaman & Nicobar Islands-744101, India*

[2]*Division of Crop Improvement and Biotechnology, Central Institute for Subtropical Horticulture, Rehmankhera, Lucknow-226101, Uttar Pradesh, India*

Introduction

Agriculture is one of the promising sectors which offers marvelous potential for sustainable growth and play a key role in the economy of the country. Many of the foods that are already common in our diet are obtained from plant varieties that were developed using conventional genetic techniques of breeding and selection. Increase in the world population and limited food supply strained researchers to introduce and advance agriculture management and modern technologies along with conventional practices to attain highest possible crop productivity. Agricultural technology is based on the domestication of wild plants to create the crops that we have come to depend upon. Humans invented agriculture approximately 10,000 years ago when they began to harvest and cultivate specific plants to produce food. The improved plant traits selected by early agriculturalists were transmitted genetically to succeeding generations of plants. Both public institutions and private companies are now involved in crop genetic improvement.

Scientists have been improving plants by changing their genetic makeup since ancient. This has been accomplished through hybridization and breeding, in which two related plants are cross fertilized and the resulting offspring have characteristics of both parent plants. In the breeding process, however, many undesirable traits often can appear in addition to the desirable ones. Some of those undesirable traits can be eliminated through additional breeding, which is time consuming. Breeders can then further select and reproduce the offspring that have the desired traits. Biotechnology is broadly defined as any technique that uses live organisms viz. bacteria, viruses, fungi, yeast, animal cells, plant cells etc. to make or modify a product, to improve plants or animals or to engineer microorganisms for specific uses (Godfray *et al.,* 2010). It encompasses genetic engineering, enzyme and protein engineering, plant and animal tissue culture technology, biosensors for biological monitoring, bioprocess and fermentation technology. Definition can include activities as diverse as making wine, beer, or bread, composting organic materials, releasing parasitic wasps to control insect pests, breeding plants and producing crops. With increasing knowledge of genetics, plant breeders in this century have accelerated the improvement of crops for enhanced yield and quality. By inserting one or more genes into a plant, scientists are able to produce a plant with new desirable trait. The new gene splicing techniques are being used to achieve many of the goals and improvements that plant breeders historically have sought through conventional methods. Biotechnology approaches have the potential to enhance crop production. New recombinant DNA techniques allows the specific identification, isolation and alteration of genes and their reintroduction into living organisms to produce transgenic varieties. These new techniques are supplementing and extending traditional breeding methods to enhance the production of food, fiber and other agricultural products. Successful applications of various biotechnological tools offer great promise to crop plants (Siddra & Iqrar, 2012).

1. Marker-assisted plant breeding

Molecular markers are identifiable DNA sequences, found at specific locations of the genome and transmitted by the standard laws of inheritance from one generation to the next. They rely on a DNA assay and in contrast to morphological markers that are based on visible traits and biochemical markers that are based on proteins produced by genes. Molecular markers are not normal genes as they usually do not have any biological effect. Instead, they can be thought of as landmarks in the genome. Markers may be either phenotypic or genotypic and marker-assisted breeding developed in the 1980s with the evolution of DNA marker technologies. Today the main DNA markers used in breeding programmes for MAS are Random Amplified Polymorphic DNA (RAPD), Amplified Fragment Length Polymorphism (AFLP), microsatellites (SSRs), SNPs and Expressed Sequence Tags (ESTs) (Table 1). Each of these markers has a different set of advantages and limits (Choudhary *et al*., 2008). Cost and possible automation of the techniques are of particular importance for their adoption. Use of molecular markers in association with linkage maps and genomics offers plant breeders the potential to make genetic progress much more precisely and rapidly than through phenotypic selection. It also offers the possibility of addressing previously unattainable goals. The success of MAS is influenced by the relationship between the markers and the genes of interest (Choudhary *et al.*, 2008; Dekkers & Hospital 2002; Dekkers 2002). There are many applications for the use of DNA markers in breeding programmes, which fall into four broad groups based on the purpose of the intervention:

a. Enhancing knowledge of breeding material and systems, such as better understanding and more effective breeding of Quantitative Trait Loci (QTL).
b. Rapid introgression or backcross breeding of simple characters as the number of back-crosses required can be reduced drastically if there are markers for the character to be introduced and for the genetic background of the recurrent parent.
c. Early or easy indirect character selection, which is important for genes that cannot be detected at an early development stage, like lysine and tryptophan genes in maize
d. New goals not possible through traditional breeding, including pyramiding of disease resistance genes with indistinguishable phenotypes.

Table 1: List of some markers used for selection of specific traits in important crop plants

Crop	Trait	Gene/QTL	Linked marker
Tomato	Yellow leaf curl virus resistance	Ty 1	RFLP
Soybean	Soybean mosaic virus resistance	rsv	Sat 309 (SSR)
Pea	Powdery mildew resistance	er	p 236 (RFLP)
Wheat	Flour colour	QTL	STS marker
Maize	Enhanced lysine and tryptophan (QPM)	opaque-2	umc 1066, phi 112
Rice	Blast resistance	Pi-1,Pi2	RZ 536, RG 64
	Bacterial blight resistance	Xa-1,3	XNpb 235, XNpb 181
Sorghum	Fertility restoration	rf4	LW 7, LW 8

2. Plant tissue culture

Tissue culture is the *in-vitro* aseptic culture of cells, tissues, organs or whole plant und controlled nutritional and environmental conditions often to produce the clones plants. The resultant clones are true-to type of the selected genotype. The controlled conditions provide the culture an environment conducive for their growth and multiplication. These conditions include proper supply of nutrients, pH medium, adequate temperature and proper gaseous and liquid environment. Plant tissue culture comprises a set of *in vitro* techniques, methods and strategies that are part of plant biotechnology. Tissue culture has been exploited to create genetic variability from which crop plants can be improved, to improve the state of health of the planted material and to increase the number of desirable germplasms available to the plant breeder. Tissue culture protocols are available for many crop species, although continued optimization is still required for many crops, especially cereals and woody plants (Thorpe; 2007; Akin-Idowu *et al.*, 2009). Plant tissue culture is considered to be the most efficient technology for crop improvement by the production of somaclonal and gametoclonal variants. The micropropagation technology has a vast potential to produce plants of superior quality, isolation of useful variants in well-adapted high yielding genotypes with better disease resistance and stress tolerance capacities (Brown and Thorpe; 1995; Garcia-Gonzales *et al.*, 2010; Sagare *et al.*, 2000). Double haploids, embryo rescue, protoplast fusion, micropropagation and somaclonal variation are important plant tissue culture techniques that have been used for improvement of crop plants.

a. Doubled haploids

Using *in vitro* techniques, it is possible to regenerate plants from pollen or ovules. These plants, which contain only one copy of each chromosome are called haploids. After appropriate chemical treatment, it is possible to restore the normal number of chromosomes and to regenerate viable plants. These plants are called double-haploids and are homozygous for all their genes. Such plants are of tremendous interest to plant breeders, since they allow development of pure line varieties or inbred parental lines much more quickly than through conventional breeding. Androgenesis (regeneration from pollen) has been successfully used for crops such as eggplant, pepper and wheat. Gynogenesis (regeneration from ovules) is used on barley. However, the bulbosum method used with barley does not require *in vitro* cultivation of ovules; development of haploids is obtained *in vivo* through interspecific crosses between barley and *Hordeum bulbosum*, a wild relative (Patnaik *et al.*, 2001; Maluszynski *et al.*, 2003; Ojaghi *et al.*, 2010).

b. Embryo rescue

Breeders need access to the largest possible genetic variability. In some cases, variability available within a given species is not sufficient to answer a specific problem. A solution available to breeders is inter-specific hybridization (crossing plants from separate but related species). However, embryos resulting from such hybridization rarely survive, due to incompatibilities between the embryo and the mother plant. Saving embryos is sometimes possible through their *in vitro* cultivation, which make it possible to isolate the inter-specific embryo from the hostile mother plant environment. This technique has been used for the introduction of disease resistance into squash, lettuce, tomato (Hossain *et al.*, 2003; Sharma *et al.*, 1996).

c. Protoplast fusion

Protoplasts are plant cells that have had their outer walls removed through chemical treatment. Protoplasts fusion is another technique that allow inter-specific hybridization between species that is impossible through conventional breeding. While it is difficult or impossible to fuse plant cells, it is possible through various techniques to merge protoplasts from different crop species or genera and then to regenerate a whole plant resulting from the fusion process. This technique has been used to introduce traits such as male sterility into rapeseed, or disease resistances in potato. Moreover, protoplast fusion is generally not effective beyond the family level due to incompatibilities between two distant genomes making it impossible to regenerate plants (Krishnaraj *et al*., 1995).

3. Transgenics

Rapid increase of human population together with global climate variability resulted in increased demand of plant based food and energy sources. Staple food grains, fruits and vegetables have essential role to enhance quality of humankind life since a diet based on cereal grains, root and tuber crops, and legumes is generally lacked a wide range of products such as fiber, vitamin, provitamins or other micronutrients and compounds exist in fruit and nut species (Varshney *et al*., 2011; Heslop-Harrison; 2005). Because an increase demand exists in global food production, many economically important field crops such as rice, maize, pulses, legumes, fruit and vegetables crops production need to be improved. However, limitation of conventional breeding due to genetic restrictions, self-incompatibility, long juvenile periods, resources restricted to parental genome and exposed to sexual combination. Thus, there is an urgent need for the biotechnology-assisted crop improvement, which ultimately aimed to obtain novel plant traits. Plant genetic engineering has opened new avenues to modify crops and provided new solutions to solve specific needs. Contrary to conventional plant breeding, this technology can integrate foreign DNA into different plant cells to produce transgenic plants with new desirable traits. These biotechnological approaches are a great option to improve plant genotypes with significant commercial properties such as increased biotic and abiotic resistance, nutrition, yield and quality and to use as bioreactor to produce proteins, edible vaccines and biodegradable plastics (Malnoy *et al*., 2010; Petri *et al*., 2011).

Techniques used for plant transform

Improvement of the plant characteristics by transfer of selected genes into plant is possible mainly through three principal methods: 1. *Agrobacterium*-mediated transformation 2. Microprojectile bombardment and 3. Floral dip. *Agrobacterium* infect a wound surface of the plants *via* Ti-plasmid. Since Ti plasmid encodes mechanisms of integration of T-DNA into the host genome, it is used as a vector to transform plants (Paszkowski & Witham, 2001). Direct gene transfer procedures involves intact cells and tissues as targets, in some species breaching of the cell wall is needed in order to enable entrance of DNA to cell. This is accomplished by making some degree of cell injury or totally enzymatic degradation of the cell wall. Microprojectile bombardment eliminates species-dependent and complex interaction between bacterium and host genome, stable integration is lower in this technique in comparison to *Agrobacterium*-mediated transformation (Taylor & Fauquet; 2002). Moreover, the existence of truncated and rearranged transgene DNA can also lead transgene silencing in the transgenic plants Advantages of microprojectile bombardment can be summarized as (a) transfer of multiple DNA fragments and plasmids with co-bombardment (b) usage of specialized vectors for DNA transfer. Application of both of the techniques for the transfer of foreign DNA results in transient or stable expression of the DNA fragment. Floral dip involves plant transformation at a high rate simply by dipping flowering plants in *Agrobacterium* that are suspended in a solution containing sucrose (or glucose) and the surfactant Silwet L-77. Plant tissue culture media, pH adjustments, plant growth regulators and the growth or use of *Agrobacterium* at particular culture densities were not essential to successful, high-rate transformation (Steven and Andrew., 1998). The great success of GM crops has had an enormous impact on world crop production and cultivation pattern of agricultural species. The extensive environmental release and cultivation of GM crop varieties have aroused tremendous biosafety concerns and debates world-wide (Stewart *et al.* 2001). Biosafety issue has already become a crucial factor in constraining the further development of transgenic biotechnology and wider application of GM products in agriculture. Presence of selectable marker genes in transgenics plants is a big issue among publics. However there are several approaches for production of clean marker-free transgenic plants.

Methods for marker free transgenic plants

a. Co-transformation
b. Multi-autotransformation
c. Site-specific recombination
d. Transposon-based marker methods
e. Chemical-inducible system
f. Heat-inducible system
g. Positive selection system
h. Negative selection system
i. Autoexcision strategy
j. Abiotic stress related gene as selection marker

4. Genomics

Genomics is analysis of complete DNA sequence of an organism. It has been one of the most transformative influences on biological studies. The genome sequences of organisms are fundamentally important for understanding the functions of individual genes and their networks, for defining evolutionary relationships, processes and for revealing previously unknown regu- latory mechanisms that coordinate the activities of genes. Hence genomics-based approaches are having a profound influence on improvement of crops for food and fuel production. Since 2005, analyses of the sequences of many crop genomes have been published (Table 2). This explosion of data has been driven by cheaper and more effective sequencing technologies (primarily the Illumina (Bentley *et al*., 2008) and Roche 454 (Margulies *et al*., 2006) methods) coupled with increasingly sophisticated sequence and assembly strategies (Maccallum *et al*., 2009), Many important agricultural traits such as yield, however, result from relatively small eûects across multiple loci. This implies that these loci might not be optimally identified through QTL approaches and that their pyramiding through MAS will be inefficient. Therefore, breeders have begun to address these problems by developing a knowledge base of associations of polymorphic markers with phenotypes in breeding populations (Morrell *et al*., 2011; Meuwissen *et al*., 2001). Here associations are used to develop a breeding model in which the frequency of desired marker alleles is optimized, thereby maximizing the estimated breeding value (Xu; 2003; Heffner *et al*., 2009). Multiple cycles of selection are used to accumulate favorable alleles that are associated with desired phenotypes. This approach, termed genomic selection (GS) is incorporated into industrial-scale breeding processes that require very cheap high-throughput marker assays (Eathington *et al*., 2007). Genomics has radically altered the scope of genetics by providing a landscape of ordered genes and their epigenetic states, access to an enormous range of genetic variation and the potential to measure gene expression directly with high precision and accuracy. This not only has important practical advantages for breeding but also facilitates systematic comparison of gene functions across sequenced genomes, bringing the wealth of knowledge of gene function and networks obtained in experimental species directly into the ambit of crop improvement (Michael and Cristobal; 2013).

Table 2: Genome size of some important crops whose genome sequence has been published with sequencing methods.

Species (common name)	Genome size	Sequence strategy	Publication year
*Oryza sativa (*rice*)*	389 Mb	BAC physical map, Sanger sequencing	2005
Populus trichocarpa	550 Mb	BAC physical map, WGS, Sanger sequencing	2006
Vitus vinifera	475 Mb	WGS, Sanger sequencing	2007
Sorghum bicolor	700 Mb	WGS, Sanger sequencing	2009
Zea mays	2,300 Mb	BAC physical map, BAC sequence	2009
Glycine max	1,115 Mb	WGS, Sange sequencing	2010
Theobroma cacao	430 Mb	WGS, Sanger, Illumina, Roche 454	2010
Brassica rapa	485 Mb	WGS, Illumina, BAC end Sanger sequencing	2011
Cajanus caja	833 Mb	WGS, Illumina	2012
Solanum lycopersicum	900 Mb	WGS, Roche 454, Illumina and SOLiD, BAC end Sanger sequencing	2012
Musa acuminate	523 Mb	WGS, Roche 454, Sanger, Illumina	2012
G. hirsutum (upland cotton)	880 Mb	WGS, Sanger, Roche 454, IlluminaIllumina	2012

5. Proteomics

The knowledge of key proteins that play crucial roles in the proper growth and development of a plant are essential to proper the biotechnological improvement of crop plants. These proteins maintain cellular homeostasis under a given environment by controlling physiological and biochemical pathways. Proteomic reveals the functional players for mediating specific cellular processes. The word proteome is derived from PROTEins expressed by a genOME. The term "proteomics" describes the study and haracterization of the complete set of proteins present at a given time in the cell (Wilkins *et al.*, 1995). The proteome, unlike the genome, which is static in nature, has dynamic capabilities. The study of proteins introduces post-translational modifications (PTMs) and provides the knowledge that is important for understanding the biological functions. The PTMs that play important roles during the growth and development of a plant and/or in response to various stress conditions cannot be understood from the genome sequence projects and/or transcript abundance alone. Proteomics knowledge provides functional genomics with completeness toward understanding the process (Park, 2004; Thurston *et al.*, 2005). Two-dimensional electrophoresis (2-DE) and mass spectroscopy (MS), two of the most widely used proteomics methods, are used to catalog and identify proteins in different proteome states or environments. Advances in 2-DE have been extremely helpful in bringing proteomics close to biotechnological programs; however, due to some drawbacks and disadvantages associated with gel-based proteomics, eg., labor intensiveness, insensitiveness to low-copy number proteins, low reproducibility and the inability to characterize complete proteomes, many gel-free proteomic techniques have also become avaluable tool for scientists (Scherp *et al.*, 2011; Jayaraman *et al.*, 2012).

The field of proteomics has been used to analyze the differences between the nutritional values of food crops through the analysis of their proteomes. Iwahashi and Hosoda (2000) reported that heat stress increased the expression of invertases in tomato fruits, thus increasing their sucrose content. As physiological disorders appears in crop if they are not harvested at right stage and may result in huge economic losses (Lliso *et al.*, 2007; Pedreschi *et al.*, 2009). Proteomic based approaches have become useful to detect biomarkers for optimal harvest maturity (Abdi *et al.*, 2002). Analysis of postharvest withering process in grapes is very critical to produce high quality wines and thus gel-based proteomics analysis of this process has been employed for improving grape quality (Di-Carli *et al.*, 2011). Also understanding the ripening and post-harvest physiology during storage will not only have impact on food quality but also on the optimization of the technological processes involved. Proteomics have investigated the reason that heat treatment for peach fruits will improve the peach fruit quality and shelf-life (Zhang *et al.*, 2011). Flour quality is highly correlated with protein composition and functional quality, thus proteomics can be very useful to identify protein markers for suitable cultivars for flour making (Yahata *et al.*, 2005). The proteomic analysis of wheat kernels for amphiphilic proteins increased the knowledge of the physiological and technological functions of wheat kernels (Amiour *et al.*, 2002). Proteomics has also helped in the construction of proteome map investigating the level of protein modification during barley malting and detecting the proteins associated with beer quality (Fasoli *et al.*, 2011).

6. Gene targeting

Gene targeting is a key technology for the rational, accurate and safe exploitation of plants through genetic manipulation. Moreover, it offers the potential to completely knockout the expression of target genes or to make specific changes to gene function, objectives that cannot be achieved by conventional transgenesis. The ability to target DNA integration would permit the locus-specific integration of a transgene into a predetermined site of the host genome, avoiding the accidental inactivation of an endogenous gene localized at the insertion site or the unexpected expression profiles of the transgene itself, the so-called position effect. Systematic isolation and sequencing of genomic DNA flanking the insertion sites (known as FSTs or Flanking Sequence Tag) offers the opportunity to rapidly characterize plants altered in a candidate gene sequence. This approach is notably most useful in fully sequenced genomes such as in *Arabidopsis thaliana*. With 125 Mbp and 26,422 genes. The recombination machinery has been well conserved throughout evolution, as an essential component of cell survival (Kumar and Sopory; 2008).

7. Metabolomics

Plants produce large numbers of metabolites of diversified structures and abundance that play important roles in plant growth, development, and response to environments. These diverse small molecular weight metabolites, the chemical base of crop yield and quality, are also valuable nutrition and energy sources for human beings and live stocks (Hall *et al.*, 2008). Generally, these metabolites are classified into primary and secondary metabolites. The former are indispensable for the growth and development of a plant, while the latter are not essential but are crucial for a plant to survive under stress conditions by maintaining a delicate balance with the environment. In addition, primary metabolites are highly conserved in their structures and abundances while those of secondary metabolites differ widely across plant kingdoms (Scossa *et al.*, 2016). The diversity of plant metabolites and the likely complicated regulatory mechanism highlight the necessity to explore the underlying biochemical nature (Hall *et al.*, 2008; Scossa *et al.*, 2016). Metabolomics has significantly contributed to the understanding of plant physiology and biology from the view of small chemical molecules that reflect the end point of biological activities, also to improve plant behavior under different conditions.

Large scale metabolite analysis of complex mixtures are carried by integrated technologies and methodologies like non-destructive nuclear magnetic resonance spectroscopy (NMR), mass spectrometry (MS) based methods including gas chromatography–MS (GC–MS), liquid chromatography–MS (LC-MS) and capillary electrophoresis–MS (CE–MS). Although metabolomics is downstream to transcriptomics and proteomics, the practical size of the metabolome of a species cannot be speculated directly by known genomic information via central dogma. Therefore, metabolomics is used to obtain a large amount of valuable information for the discovery of genes and pathways. With the growing interest in the use of metabolomic technologies for a wide range of biological targets, plant metabolomics have dramatically improved in recent years. Plant metabolomics alone or combined with functional genomics has been applied in many fields. Metabolomics provides an additional dimension to GM crop analysis, allowing the detection of both intended and unintended effects that might take place in GM crops because of genetic modification at the metabolic level. This facilitates greatly the substantial equivalence evaluation of GM crops. Case studies comparing metabolic changes between GM crops and their non-transgenic counterparts have covered almost all important agricultural crops, such as rice, maize, soybean, pea, wheat, potato, tomato, barely, which have confurmed detectable alterations of their metabolites due to transgenic modifications (Simó *et al.*, 2014; Chang *et al.*, 2012).

7. Genome editing

It is a type of genetic engineering in which DNA is inserted, deleted or replaced in the genome of a living organism using engineered nucleases or molecular scissors. These nucleases create site-specific double-strand breaks (DSBs) at desired locations in the genome. The induced double-strand breaks are repaired through nonhomologous end-joining (NHEJ) or homologous recombination (HR), resulting in targeted mutations. To create targeted mutations, sequence-specific nucleases (SSNs) must recognize a specific DNA sequence signature within a complex genome. There are currently four major classes of sequence-specific nucleases (SSNs) that can accomplish by zinc finger nucleases (ZFNs), transcription activator-like effector nucleases (TALENs), meganucleases, and clustered regularly-interspaced short palindromic repeats (CRISPR) and CRISPR-associated (Cas) nuclease 9 (CRISPR/Cas9). Zinc-finger nucleases (ZFNs) are targetable DNA cleavage reagents that have been adopted as gene-targeting tools. ZFN-induced double-strand breaks are subject to cellular DNA repair processes that lead to both targeted mutagenesis and targeted gene replacement at remarkably high frequencies. Meganucleases are naturally occurring restriction enzymes that recognize and cleave DNA sequence targets, typically from 12 to 40 bp (Paques and Duchateau, 2007). Meganucleases can be engineered to recognize new sites; however, changes in target site specificity often result in a reduction of catalytic activity. Meganucleases, ZFNs and TALENs all recognize their sequence targets through protein/DNA interactions, while CRISPR/Cas9 targeting is achieved through a guide RNA (gRNA) that base pairs with the chromosomal target of interest (Mali et al., 2013). The Cas9 nuclease then cleaves the resulting RNA/DNA complex. The simplicity of DNA targeting through base-pairing has led to the quick and broad adoption of CRIPSR/Cas9 reagents for genome editing. It is now possible to create locus-specific DNA DSB with ease. Genome editing has been successfully utilised for gene correction/addition in *Arabidopsis* (ADH1, TT4), tobacco (chitinase) and maize (zeatin15).

Conclusion

The improvement of agricultural production and productivity as well as the future versatility of agricultural production are dependent on the rational utilization of technologies. Biotechnology approaches have the potential to enhance crop production. Plant tissue culture techniques is major industrial importance in the area of plant propagation, disease elimination, plant improvement and production of secondary metabolites. The micropropagation technology has a vast potential to produce plants of superior quality, isolation of useful variants in well-adapted high yielding genotypes with better disease resistance and stress tolerance capacities. The future of Plant transformation is required of genotype-independent protocols, efficiency and highly-yield selection and regeneration *in vitro* culture methodology. Genomics and proteomics based approaches are having a profound influence on improvement of crops for food and fuel production. Genome sequences of organisms are important for understanding the functions of individual genes and their networks, for defining evolutionary relationships, processes and for revealing previously unknown regulatory mechanisms that coordinate the activities of genes. The knowledge of key proteins which are associated with cellular homeostasis under a given environment and plays role in proper growth and development of a plant are critical to proper the biotechnological improvement of crop plants. Fast emergence of technologies such as miRNA, targeted gene replacement using zinc-finger nucleases, chromosome engineering and nanobiotechnology the future seems bright with respect to the development of designer crops with improved features that can use natural resources such as water, soil nutrients with a far greater efficiency for enhancing the production and productivity. Plant biotechnology and genomics presents considerable improvement in almost every area of crop production with possible profit for farmers, industries and consumers. The growth in the world population, their demands for food and clear consumer preference for environmentally sustainable agriculture will extend biotechnology's role in food production. Successful applications of various biotechnological tools offer great promise to crop plants in future.

References

Abdi N. Holford P. Mcglasson B (2002). Application of two-dimensional gel electrophoresis to detect proteins associated with harvest maturity in stonefruit. *Postharvest Biol. Technol.* 26: 1–13

Akin-Idowu PE. Ibitoye DO. Ademoyegun OT (2009). Tissue culture as a plant production technique for horticultural crops. *Afr. J. Biotechnol.* 8(16): 3782-3788.

Amiour N. Merlino M. Leroy P. Branlard G (2002). Proteomic analysis of amphiphilic proteins of hexaploid wheat kernels. *Proteomics* 2: 632–641.

Bentley DR. Balasubramanian S. Swerdlow HP. Smith GP. Milton J. Brown CG. Hall KP. Evers DJ. Barnes CL. Bignell HR. Boutell JM. Bryant J. Carter RJ. Keira Cheetham R. Cox AJ. Ellis DJ. Flatbush MR. Gormley NA. Humphray SJ. Irving LJ. Karbelashvili MS. Kirk SM. Li H. Liu X. Maisinger KS. Murray LJ. Obradovic B. Ost T. Parkinson ML. Pratt MR (2008). Accurate whole human genome sequencing using reversible terminator chemistry. *Nature*, 456:53-59.

Brown DCW. Thorpe TA (1995). Crop improvement through tissue culture. *World J. Microbiol & Biotechnol*. 11:409-415.

Chang Y. Zhao C. Zhu Z. Wu Z. Zhou J. Zhao Y. Lu X. Xu G. (2012). Metabolic profiling based on LC/MS toevaluate unintended effects of transgenic rice with cry1Ac and sck genes. *Plant Mol. Biol.*, 78: 477–487.

Choudhary K. Choudhary OP. and Shekhawat NS (2008) Marker Assisted Selection: A Novel Approach for Crop Improvement . *American-Eurasian Journal of Agronomy* 1(2): 26-30.

Dekkers JCM (2004). Commercial application of marker-and gene-assisted selection in livestock: strategies and lessons. *J. Animal Sci.*, 82: E313-E328.

Dekkers JCM and Hospital F (2002). The use of molecular genetics in the improvement of agricultural populations. *Nature Reviews-Genetics*, 3: 22-32.

Di-Carli M. Zamboni A. Pezzotti M. Lilley KS. Benvenuto E (2011). Two-dimensional differential in gel electrophoresis (2D-DIGE) analysis of grape berry proteome during postharvest withering. *J. Proteome Res.* 10, 429–446.

Eathington SR. Crosbie TM. Edwards MD. Reiter RS. Bull JK (2007). Molecular markers in a commercial breeding program. *Crop Sci.*, 47:S154-S163.

Fasoli E. D'Amato A. Kravchuk AV. Citterio A. Righetti PG (2011). In-depth proteomic analysisof non-alcoholic beverages with peptide ligand libraries. I: almond milk and orgeat syrup. *J. Proteomics* 74:1080–1090.

Garcia-Gonzales R. Quiroz K. Carrasco B. Caligari P (2010). Plant tissue culture: Current status, opportunities and challenges. *Cien. Inv. Agr*. 37(3): 5-30.

Godfray HCJ. Beddington JR. Crute IR. Haddad L. Lawrence D. Muir JF. Robinson S. Thomas SM (2010). Food Security: the challenge of feeding 9 billion people. *Science,* 327: 812–818.

Hall R.D. Brouwe I.D. Fitzgerald M.A. (2008). Plant metabolomics and its potential application for human nutrition. *Physiol. Plant.*, 132: 162–175.

Heffner EL. Sorrells ME. Jannink JL (2009). Genomic selection for crop improvement. *Crop Sci.*, 49:1.

Iwahashi Y. Hosoda H (2000). Effect of heat stress on tomato fruit protein expression. *Electrophoresis* 21: 1766–1771.

Jayaraman D. Forshey K L. Grimsrud PA. Ane J M (2012). Leveraging proteomics to understand plant–microbe interactions. *Front. Plant, Sci.* 3:44.

Kumar, A. and Sopory, S. (2008): Recent Advances in Plant Biotechnology and its Applications. New Delhi. I.K. International, 718 pp.

Krishnaraj S. & Vasil IK (1995) Somatic embryogenesis in herbaceous monocots. In: *in vitro Embryogenesis in Plants*, ed Thorpe, T.A. 417–471. Dordrecht: Kluwer Academic. Litz). CAB International, London, UK.

Lliso I. Tadeo FR. Phinney BS. Wilkerson CG. Talón M (2007). Protein changes in the albedo of citrus fruits on postharvesting storage. *J. Agric. Food Chem.*, 55, 9047–9053.

Maccallum I. Przybylski D. Gnerre S. Burton J. Shlyakhter I. Gnirke A. Malek J. McKernan K. Ranade S. Shea TP. Williams L. Young S. Nusbaum C. Jafe DB (2009). ALLPATHS 2: small genomes assembled accurately and with high continuity from short paired reads. *Genome Biol.* 10:R103.

Mali P. Yang L. Esvelt K. M. Aach, J. Guell M. DiCarlo J.E. Norville J. E. Church G. M. (2013). RNA-guidedhuman genome engineering via Cas9. *Science* 339: 823–826.

Malnoy M. Boresjza EE. Norelli JL. Flaishman MA. Gidoni D. Aldwinckle HS (2010). Genetic transformation of apple (Malus x domestica) without use of a selectable marker gene. *Tree Genetics and Genome*, 6: 423-433.

Maluszynski M. Kasha KJ. Forster BP. Szarejko I (2003). Doubled haploid production in crop plants: A manual. Kluwer Academic Publ, Dordrecht, Boston, London, 1-20.

Margulies M. Egholm M. Altman WE. Attiya S. Bader JS. Bemben LA. Berka J. Braverman MS. Chen Y-J. Chen Z. Dewell SB. Du L. Fierro JM. Gomes XV. Godwin BC. He W. Helgesen S. Ho CH. Irzyk GP. Jando SC. Alenquer MLI. Jarvie TP. Jirage KB. Kim J-B. Knight JR. Lanza JR. Leamon JH. Lefkowitz SM. Lei M. Li J. (2006). Genome sequencing in microfabricated high-density picolitre reactors. *Nature*, 441:120.

Meuwissen TH Hayes BJ Goddard ME (2001). Prediction of total genetic valueusing genome-wide dense marker maps. *Genetics*, 157:1819-1829.

Michael WB. Cristobal U (2013). Genomics reveals new landscapes for cropimprovement. *Genome Biology* 14:206-218.

Morrell PL. Buckler ES. Ross-Ibarra J (2011). Crop genomics: advances and applications. *Nat Rev Genet.*, 13:85-96.

Ojaghi J and Akhundova E (2010). Genetic diversity in doubled haploids wheat based on morphological traits, gliadin protein patterns and RAPD markers. *African J Agric Res*, 5:1701-1712.

Paques F. and Duchateau P. (2007). Meganucleases and DNA double-strand break-induced recombination: perspectives for gene therapy. *Curr. Gene Ther.* 7:49–66

Park O K (2004). Proteomic studies in plants. *J. Biochem. Mol. Biol.*133–138.

Paszkowski J. Witham SA (2001). Gene silencing and methylation processes. *Current Opinion in Plant Biology*, 4:123–129.

Patnaik D. & Khurana P (2001). Wheat Biotechnology: A mini review. *Elecc J Biotech*, 4: 1-29.

Pedreschi R. Hertog M. Robben J. Lilley KS. Karp NA. Baggerman G (2009). Gel-based proteomics approach to the study of metabolic changes in pear tissue during storage. *J. Agric. Food Chem*. 57: 6997–7004.

Petri C. Hily J.M. Vann C. Dardick, C. Scorza R (2011 A high-throughput transformation system allows the regeneration of marker-free plum plants (*Prunus domestica*). *Annals of Applied Biology*, 159: 302-315.

Sagare AP. Lee YL. Lin TC. Chen CC. Tsay HS (2000). Cytokinin-induced somatic embryogenesis and plant regeneration in Coryodalis yanhusuo (Fumariaceae)- a medicinal plant. *Plant Sci.* 160: 139-147.

Scherp P. Ku G. Coleman L. Kheterpal I (2011). Gel based and gel-free proteomic tech-nologies. *Methods Mol. Biol.* 702,163–190.

Scossa F. Brotman Y. de Abreu e Lima F. Willmitzer L. Nikoloski Z. Tonga T. Fernie A.R. (2016). Genomics-based strategies for the use of natural variation in the improvement of crop metabolism. *Plant Sci.*, 242: 47–64.

Sharma D.R. Kaur R and Kumar (1996). Embryo rescue in plants—a review *Euphytica*. 89, Issue 3:325-337.

Siddra I. Iqrar AR (2012). An Overview of Biotechnological Approaches for Crop Plant Improvement *Genomics and Applied Biology*, 3 (3): 22-26.

Simó C. Ibáez, C. Valdés A. Cifuentes A. García-Cañas V. (2014). Metabolomics of genetically modified crops. *Int. J. Mol. Sci.*, 15: 18941–18966.

Stewart CN. (2001). The utility of green fluorescent protein in transgenic plants. *Plant Cell Reports*, 20: 376–382.

Taylor NJ. Fauquet CM (2002). Microparticle Bombardment as a Tool in Plant Science and Agricultural Biotechnology. *DNA and Cell Biology*, 21: 963–977.

Thorpe T (2007). History of plant tissue culture. *J. Mol. Microbial Biotechnol.* 37: 169-180.

Thurston G. Regan S. Rampitsch C. Xing T (2005). Proteomic and phosphoproteomic pproaches to understand plant–pathogen interactions. *Physiol. Mol. Plant Pathol.* 66: 3–11.

Varshney RK. Kailash C. Bansal KC. Aggarwal PK. Datta SK. Craufurd PQ (2011). Agricultural biotechnology for crop improvement in a variable climate: hope or hype? *Trends in Plant Science*, 16: 363-371.

Wilkins MR. Sanchez JC. Gooley AA. Appel RD. Humphery-Smith I. Hochstrasser DF (1995). Progresswith proteome projects: why all proteins expressed by a genome should be identified and how to do it. Biotechnol. *Genet. Eng. Rev.* 13:19–50.

Xu SS (2003). Estimating polygenic effects using markers of the entire genome. *Genetics*, 163: 789-801.

Zhang L. Yu Z. Jiang L. Jiang J. Luo H. Fu L (2011). Effect of post-harvest heat treatment on proteome change of peach fruit during ripening. *J. Proteomics* 74, 1135–1149.

4

Breeding for Abiotic Stress Tolerance in Vegetable Crops: Challenges and Opportunities

Patel Sayeed Abdul Hamid, Ajmer Singh Dhatt
Sat Pal Sharma and Prerna Thakur

Department of Vegetable Science, Punjab Agricultural University
Ludhiana-141 004, India

Abstract

Abiotic stresses have been a serious challenge for sustainable agriculture particularly vegetable cultivation characterized by decreased endurance against stresses like drought, temperature fluctuation and salinity. In India, drought and salinity being major abiotic stresses affect an area of 16% and 4.28%, respectively. Plant Breeders have to face many impediments while developing stress tolerant varieties. Developing tolerant varieties is quite difficult as tolerance varies with the developmental stages and has complex inheritance. This can be understood well by studying ontogeny of plants. Drought varies in time of occurrence, severity and duration. Field screening for tolerance is very difficult because of various interactions like drought × high temperature, high temperature × salinity interactions etc. Artificial screening in different stresses could be an alternative approach. The varieties developed with higher inputs at research stations fail to perform at farmers' fields due to high genotype × environment interactions. Cultivated species have less genetic variability for abiotic stress tolerance but tolerance can be enhanced by using diverse and heterogeneous population. Though wild species possess genes for tolerance but these genes are closely linked with the undesirable gene(s) e.g. in tomato where S. cheesmanii, S. peruvianum, S. pimpinellifolium and S. sitiens have undesirable traits which affect yield and quality adversely. Further, the crossibility barriers, in crosses of Abelmoschus esculentus × A. ficulneus, Capsicum annuum × C.

baccatum, is another hindrance. Further, quantitative traits have low heritability and low genetic gain per selection cycle. Two approaches have been proposed for improving crop stress tolerance, one by utilizing genetic variation through direct or biotechnological approaches, and second by creating variability with transgenics, mutation breeding or protoplast fusion. Currently, equipped with advanced breeding techniques like genomics, transcriptomics, proteomics and metabolomics, vegetable breeders are able to develop stress tolerant varieties. Transgenic technology will also broaden the sources of genes for tolerance across different living organisms. A closely knitted multidisciplinary approach is the call of future for breeding abiotic stress tolerant varieties in vegetable crops.

***Keywords**: Abiotic stress, vegetables, drought, salinity, genotype × environment interactions*

Introduction

Changing climate has created serious challenges and put unprecedented pressure on sustainable agriculture production. Vegetable farming has been crippled by the inherent nature of vegetable itself *i.e.* succulence. This further subside their endurance towards an array of abiotic stresses like drought, temperature fluctuation and salinity. It is estimated that crop grown on 90 per cent of arable land experiences one or more environmental stresses (Latef *et al.*, 2015). These stresses may translate into losses to the tune of 70 per cent in yield of vegetable crops (Shrivastava and Kumar, 2015). The cumulative effects of multiple stresses may reduce fruit yield, post harvest and nutrient quality. In India, drought and salinity are the major abiotic stresses and they encroaches an area of 16 per cent and 4.28 per cent, respectively (Anonymous 2013, 2011).

By the end of 21st century, a predicted rise in temperature was 1.5-5.8°C which further intensify the stresses and creates great nuisance to agriculture production (Rosenzweig *et al.*, 2001). Various physiological disorders were associated with abiotic stresses in vegetable crop which are mostly mystified with nutritional deficiency. Some abiotic disorders were mentioned in table 1.

Table 1: Abiotic disorders in vegetable crops

Crop	Disorder	Cause
Onion	Bolting	Frost injury (Low temperature during bulbing)
	Splitting	Water stress followed by heavy irrigation
Cabbage	Browning/brown rot/red rot	Salinity
	Loose/puffy head	Temperature fluctuation
	Pepper spot/black speck	Temperature fluctuation along with high dose of fertilizer
Cauliflower	Blindness	Subzero temp within few days of transplanting along with physical damage
	Buttoning	Poor soil moisture and low temperature
	Leafyness/bracting	High temperature during cure development
	Premature bolting	Temperature fluctuation
	Riceyness	High temperature beyond optimum range
Cucurbits	Frost/low temperature injury	Sudden fall in temperature down to sub zero
	Internal rind necrosis (Watermelon)	Moisture stress
	Leaf silvering (Pumpkin)	Drought Stress along with whitefly
Beetroot	Zoning	High temperature
Potato	Tuber cracking and internal spot	High temperature
	Jelly end/ sugar end	Low soil moisture and high temperature during early tuber growth
Chilli	Stip disorder	High temperature along with other abiotic stress
Tomato	Blossom drop	High temperature and dry weather
	Blotchy ripening / gray wall	High temperature along with nutritional deficiency during ripening
	Cat face	Low temperature (<10°C) at pollination and fruit setting
	Internal white tissue	High temperature (>32°C) during ripening along with potash deficiency

Naturally plant adapts some strategies for its survival under stressful conditions. For example under drought stress, plant complete it life cycle before the occurrence of drought (drought escape) or plant able to maintain water balance (drought avoidance) or plant yield higher even under low water potential (drought tolerance). Under saline condition, plant selectively absorb or exclude mechanisms and glands/vacuoles which secret salt and behave like halophytes (salt avoidance) while even after absorbing salt ions, plant able to yield significantly by maintaining turgor pressure or tissue dehydration tolerance under stress condition (non-halophytes). Under saline condition, exclusion of salt from leaf tissue, stem or accumulation of salt in leaf tip (garlic) etc was observed in salt tolerant vegetable crops. Lunin *et al.* (1963) proposed that the actual tolerance vary with growth stages at which salinization is initiated and the final level of salinity achieved and salt tolerance should be considered with marketable yield. Increase in vegetable production can be achieved through development of varieties tolerant to abiotic stress. Breeding cultivars with enhanced environmental stress tolerance is one of the important strategies to manage the associated crop losses (Mutava *et al.* 2015).

This chapter emphasize mainly on the challenges faced by the breeder and common breeding methods followed for developing abiotic stress tolerance. It also focused the ways to solve that challenges by the use of various breeding and advanced tools which includes QTL mapping, marker assisted selection, genomic selection and transgenic approaches.

Breeding challenges

Even though such above discussed strategies were adapted by plant, the process is slow and not economically feasible to breeder as well as farmer. Among challenges, breeder face as many as hindrance for developing tolerant cultivars. Among them, few are most important and faced by all breeder community worldwide.

A) Developmental stages and complex nature

Many researchers reported that tolerance to abiotic stress vary with the species, genotype as well as within genotype itself also. Cuartero *et al.* (2006) stated that tolerance at one stage of development is not often correlated with the tolerance at another stage. This also denotes that different genetic and physiological mechanisms are contributing at different stages (Pineda *et al.* 2012). For successful crop production under stressful environment, tolerance must be at all major plant development stages, right from seed germination, vegetative stage, flowering to fruit/ crop maturity (Foolad, 2007).

In table 2, with same mapping population i.e. BC_1F_1, Foolad (2007) found different QTLs at different developmental stages in tomato for salt tolerance. Even he reported that different QTLs were expressed within same stages. So we can conclude that, many minor QTLs are controlling abiotic stress tolerance which makes it fate as complex nature. Even if major QTLs (30%) were

Table 2 : QTL for salinity stress in tomato

Stress	Stage	QTLs	Tolerance source	Mapping population	Chromosome location
Salt	Seedling	5	*Solanum pennellii*	F_2	1,3,7,8,12
		8	*S. pennellii*	F_2	1,2,3,7,8,9,12
		8	*S. pennellii*	F_2	1,3,5,6,8,9
		7	*S. pimpinellifolium*	BC_1F_1	1,2,5,7,9,12
	Vegetative	8	*S. pimpinellifolium*	BC_1F_1	1,2,3,4,8,9,12
		4	*S. pimpinellifolium*	RILs	1,5,9
		5	*S. pimpinellifolium*	BC_1	1,3,5,6,9
		7	*S. pimpinellifolium*	RILs	1,3,4,5,7,8,9

transferred, we can't achieve complete tolerance. At different developmental stages or under abiotic stress conditions, a DNA methylations, histone modification and small RNA patterns may also change the expression or response of plant under stress condition. These changes could be epigenetic and can occur genome wide in *Arabidopsis* (Cokus *et al.* 2008; Roudier *et al.* 2011). For such studies, identification of tolerant and susceptible accessions must be done through stringent screening.

B) Difficulty in screening

Each developmental stage may require a different screening procedure and simultaneous or sequential screening may be impractical or impossible. However, quantification of tolerance often poses serious difficulties. This may be because under natural environment, soil and atmospheric condition varies from site to site and year to year (Hall, 2011). Further stresses get interacted at both plant and cellular level, involving interaction of stress with molecular, biochemical and physiological processes (Ashraf *et al.,* 2008). With the interaction between them, i.e. drought × high temperature, high temperature × salinity *etc.* further catalyzes oxidative and osmotic stress which makes breeder to identify tolerant plant under natural condition against particular stress.

Selection under natural condition is being traditionally followed widely by the breeder but the results are significantly different in replications and year data also. In a traditional breeding process, breeders primarily select germplasm to cross with an elite cultivar and select a new line from the offspring based on their phenotypic performances. Because of the complex nature of abiotic stress tolerance, it is extremely difficult to improve this trait using only the empirical method of breeding.

For salinity, germination test (Carlson *et al.,* 1983), sand culture (Sharma and Kumar, 1985), hydroponics (Jana and Slinkard, 1979), pollen grain screening (Sacher and Mulcahy, 1981) and young seedling screening (Datta and Pradhan, 1981) etc different screening methods were reported but not commonly used. For drought tolerance screening, except natural screening, polyethylene glycol (Kaufmann, 1969) and wooden box method (Aamodt and Johnston, 1936) are also being used. While for temperature fluctuation hot bed method is also reported to screen for high temperature tolerance. Some indirect selection methods include membrane stability, chlorophyll content, pollen viability, vigor index etc are also been used to screen for abiotic stress tolerance. Some of the tolerant species or cultivars were identified for abiotic stress in vegetable crops (Table 3).

Table 3 : Sources of tolerance for abiotic stress in vegetable crops

Crop	Species/Cultivars
Drought and high temperature stress	
Tomato	*Solanum habrochaites* (EC-520061), *S. pimpinellofolium* (PI-205009, EC-65992, Pan American), *S. esculentum* var. *cerasiforme, S. hirsutum, S. cheesmanii, S. chilense. S. pennellii, S. peruvianum, S. sitiens* (syn *S. ricikii*)
Brinjal	*Solanum macrocarpon, S. gilo, S. macrosperma, S. integrifolium,* Supreme, Violette, PKM-1
Chilli	*Capsicum chinense, C. baccatum* var. *pendulum, C. eximium*, Arka Lohit
Potato	*Solanum acaule, S. demissum, S. stenotomum, S chacoense, S. ajanhuiri,* Alpha, *S. kurzianum, S. cardiophyllum* subsp. *Ehrenbergii*
Pea	*Pisum acutifolius*
Onion	*Allium fistulosus, A. munzii,* Arka Kalyan, MST 42
Okra	*Abelmoschus caillei, A. rugosus, A. tuberosus*
Cucumber	*Cucumis pubescens,* INGR-98018
Carrot	Pusa Kesar, Pusa Vasuda, Pusa Vrishti (High temperature)
Water melon	*Citrullus colocynthis*
Muskmelon	*Cucumis melo* var *momordica, C. melo.* var *callosus*
French bean	*Phaseolus acutifolius* (PI 321 638)
Sweet potato	*Ipomoea triloba, I. lacunose*, VLS-6, IGSP 10, IGSP 14, Sree Bhadra
Salinity stress	
Tomato	*Solanum pimpinellofolium, S. cheesmanii, S. chilense. S. pennellii, S. peruvianum, S. sitiens, S. galapagense*
Brinjal	*Solanum integrifolium*
Potato	*Solanum kurzianum*
Pea	*Pisum filiformis*
Onion	Hisar-2
Cucumber	*Cucumis pubescens*
Lettuce	*Lactuca. serriola, L. vignata, L. saligna*
Frost stress	
Tomato	*Solanum habrochaites*
Potato	*Solanum acaule, S. megistacrolobum, S. commersonii*, Kufri anand, Kufri arun, *S. ajanhuiri, S. × juzeoczukii, S.* x *curtilobum*
Pea	*Pisum filiformis, P. angustissimus*
French bean	*Phaseolus coccineus*
Okra	*Abelmoschus angulosus*
Sweet potato	*Ipomoea lacunose*

Hypothetical G×E interaction of cross-over type between experiment station and farmers fields.

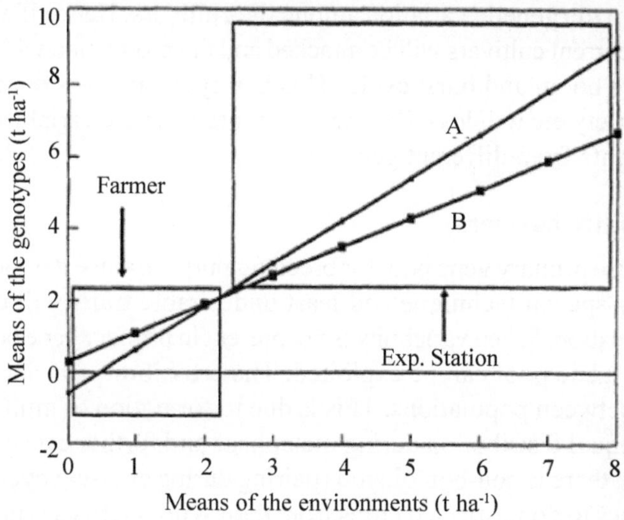

Fig. 1 : Depict that, under two different geographical areas, same genotype will perform differently. Management practice and climatic condition vary from place to place and hence genotype developed at experimental station may not perform well under farmers' field i.e. under low input management practice which nullifies the efforts of breeder. This G × E interaction can be curtled down by following ways:

- Direct selection for specific adaptation in the target environment
- Use of locally adapted germplasm *i.e.* wild relatives, land races and farmers variety.
- Test sites representing target environment or identifying/creating hot spot as like target environment.

C) Low genetic variability

One of the main reasons for high G×E interaction is low genetic variability. This may be because of two main reasons: a) hybrid technology and b) vulnerable to new biotic and abiotic stresses. From the year 1971, release of 1st F_1 hybrid i.e. Pusa Meghdoot in bottle gourd by IARI, New Delhi and Karnataka hybrid in tomato by IAHS, Bangalore, etc hybrid replaced most of the landraces, farmer varieties, etc. the genetic variability was mostly accessed by biochemical or molecular marker. Although the loss of variability is largely due to replacement of landraces by modern varieties, population pressure, urbanization and environmental degradation. Later reason could be due to genetic wipeout or genetic erosion. With the use of high dose biocides, new strains of pathotypes are developing with high level of resistance.

The genetic erosion could be due to use of single source of resistance or single parent based hybrids. This leads to have same genetic make-up of hybrid and narrow down the genetic variation among the cultivars. Hence if new pathotype evolve, the current cultivars will be attacked and the production will fall suddenly. This follows boom and burst cycle. This is a cyclic process once a landraces, farmers' variety etc will loss. But the way out to increase variability is by using diverse parents from different gene pool.

D) Crossibility barriers

Mostly we use primary gene pool for breeding purpose since it is easily crossable without any special technique and least undesirable traits will come into the cross combination. When variability is not present in primary gene pool, secondary and tertiary gene pool can be exploited. The crossibility barriers obstruct the gene flow between populations. This is due to formation of multivalent which leads to unequal distribution during metaphase and further cause F_1 sterility. In other words, there is non-homologous pairing during crossing over. Transferring tolerant genes to domesticated crops from their wild relatives is not an easy task due to some reproductive barriers (Ashraf *et al.,* 2008). But for using secondary and tertiary gene pool, special technique is required like tissue culture, ovule culture, bridging species, use of mixed pollen, washing stigma with organic solvent (n hexane), etc.

Table 4: Crossibility relationship with some wild species of respective vegetable crops

Crop	Cross incompatible	Other technique
Chilli	*C. annuum* × *C. baccatum*	-
	C. annuum × *C. pubescens*	-
	C. annuum × *C. pendulum*	Embryo culture
	C. pubescens × *C. chinense*	"
Tomato	(*S. lycopersicum* × *S. pennellii*)	Bridging sp.
	S. sitiens S. lycopersicum × *S. peruvianum*	-
Brinjal	S. melongena × *S sisymbriifolium*	Embryo culture
Potato	(*S. demissum* × *S. rybinii*) *S. tuberosum*	Bridging sp.
	(*S. acaule* × *S. simplicifolium*) *S. tuberosum*	"
Okra	*A. esculentus* × *A. ficulneus*	Ovule culture
Pumpkin and Squashs	*Cucurbita moschata* × *C. ficifolia*	Embryo culture
	C. mixta × *C. ficifolia*	"
Melon	*Cucumis figarei* × *C. melo*	Washing stigma with organic solvent
Amaranthus	*A. caudatus* × *A. hypochondriacus*	Embryo culture
	A. edulis × *A. hypochondriacus*	"

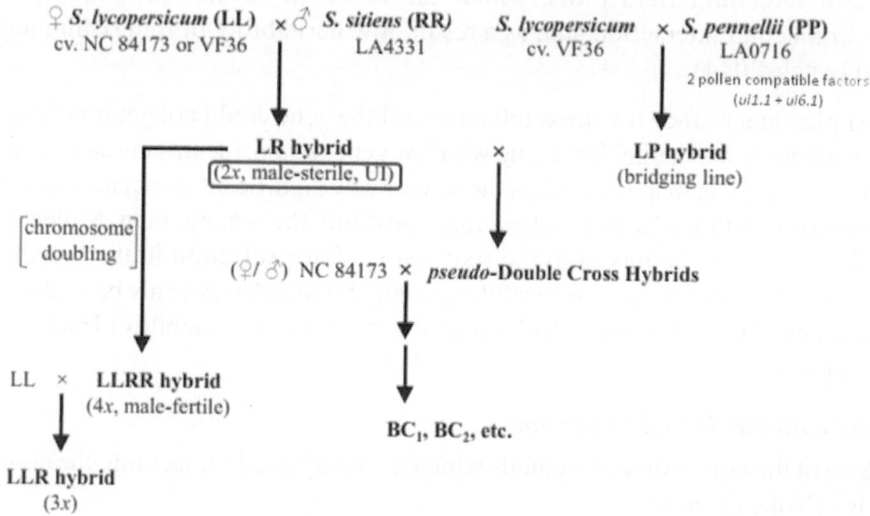

Fig 2: Crosses used to synthesize diploid, allotetraploid and allotriploid *S. lycopersicum* × *S. sitiens* hybrids

By screening the available germplasm stock including cultivated and wild species, such tolerant genotypes can be involved in breeding programme. By the use of above mentioned bridging techniques, a tolerance can be transferred from wild relatives in to a cultivated background and can be multiplied with tissue culture technique (Table 4). Chetelat (2015) overcame sterility and unilateral incompatibility by using bridging species i.e. *S. penellii* (PP) between *S. Lycopersicum* (LL) and *S. sitiens* (RR). Fig 2 shows that, LL × RR is incompatible and hence produced sterile F_1. This sterile F_1 was doubled (LLRR; allotetraploid) and crossed with LL and allotriploid was obtained. But this cannot be used further so PP was used as bridging species which have two pollen compatible factors and the F_1 (bridging line) was crossed with RR and fertile F_1 was obtained which was further backcrossed with LL to obtain abiotic tolerance of RR in a background of LL. So such techniques can be used to overcome crossibility barriers.

E) Low heritability and genetic gain

Even if such tolerant species from secondary and tertiary gene pool were identified and used in breeding programme, some undesirable traits must get transferred which are tightly linked to tolerant gene. Most probably it is due to the intensity of stress which varies from year to year and thus unavoidable environmental selection pressure on breeding material changes drastically from generation to generation (Kumar *et al.* 2012) which leads to slow breeding activities and complicated (Roy and Murty, 1970). Spatiotemporal variability within screening field plots, which can result in substantial genotype × environmental interaction and, as a result, low heritability of trait (Chimungu and Lynch, 2014).

Complex inheritance for stress tolerance, all the gene could not get transferred which leads to low heritability along with low genetic gain. Identifying and placing QTL on a genetic map is based on the statistically significant association of trait expression with molecular makers that constitute the genetic map. Molecular makers found to be linked to the trait can used for selection in the breeding process. Molecular makers are useful especially for trait that can only be evaluated on mature plants and with high plasticity resulting in low heritability (Trachsel *et al.,* 2010).

Conventional breeding method

Some of the conventional methods which are being used for tackling the above listed challenges are:

- Introduction and selection
- Wide hybridization
- Backcross breeding
- Mutation breeding
- Recurrent selection

Among these methods, Introduction followed by selection is most common breeding method. The variety which is developed at different station was tested at their location and accordingly released directly or release after few selection cycles. Wide hybridization has its own limitation of transferring undesirable traits with the desirable once (Dauney, 2008). Still some of the achievements are reported in literature. Regarding backcross breeding, this method is mostly used when a trait is governed by monogene. But to transfer major QTL in a desirable background, this method was also used along with molecular marker (Goodstal *et al.,* 2005).

The main problem conventional plant breeders have been facing is the low magnitude of genetically based variation in the gene pools of most crop species (Ashraf and Akram, 2009). Since the variability is less in cultivated species, it can be created by using mutation breeding. But the chance of getting desirable mutant is very less (10^{-6}) and it is a hit and try method as it may alter the normal gene and cause undesirable epigenetic change (Donald, 1968). González *et al.* (2005) found two promising mutant i.e. M4 and M15 (Table 5) after irradiating with ^{60}Co γ ray at dose of 300 and 500 Gy on INCA-9-1 and Amalia parent respectively. These two mutants are significantly different from other mutant and they even perform very well in different region of the country.

Table 5: Induction of drought tolerance in tomato using ^{60}Co Gamma ray irradiation

Variety/ mutant	Treatment (Gy)	Yield/ plant (kg)	Fruit number/ plant	Average fruit weight (g)
INCA-9-1	Parent	1.36 ± 0.15	32.70 ± 2.9	41.60 ± 1.9
M19	500	2.69 ± 0.14	39.30 ± 1.8	68.59 ± 0.5
M4	300	2.72 ± 0.14	33.10 ± 1.5	82.15 ± 1.2
M20	300	2.16 ± 0.04	53.10 ± 1.0	40.71 ± 1.1
M16	300	2.05 ± 0.03	46.10 ± 2.5	44.64 ± 1.5
AMALIA	Parent	1.41 ± 0.17	17.60 ± 1.0	88.45 ± 2.0
M15	500	3.69 ± 0.09	46.50 ± 2.9	79.84 ± 3.8
M17	500	2.34 ± 0.12	23.30 ± 4.3	106.22 ± 8.3
M2	300	2.51 ± 0.11	26.60 ± 1.0	94.02 ± 4.7

Abiotic stress tolerance can be incorporated into new cultivated variety by using distinct parent in every recurrent selection cycle. There is a gradual accumulation of desirable alleles per selection cycle. It differs from other as it size of population and simultaneously increases variability and genetic gain per selection cycle (Doná *et al.*, 2012). Inbred lines can also be extracted from the hybrid as there is continuous recombination because of open pollination (Hallauer, 1985). Benites and Pinto (2011) estimated the genetic gains of three cycles of recurrent selection at two growing season for heat tolerance in potato. The genetic gains for tuber yield were 37.85% followed by 32.40% for per cent large tuber and 12.99% for tuber weight over base population after three selection cycles (Table 6). So, recurrent selection method can be efficiently utilized for breeding abiotic stress tolerance cultivar in other vegetable as well. But the main disadvantage of this method is long selection cycle. The above all obstacles can be avoided or passed by use of advance tool.

Table 6 : Genetic gains for heat tolerance in potato in three cycles of recurrent selection (RS)

Cycle of RS	Tuber Yield	Tuber weight	% large tuber	Specific gravity
Base Population	558.01	117.76	41.88	1.07
Cycle-I	1170.72	147.14	76.00	1.08
Cycle-II	990.95	150.86	76.25	1.09
Cycle-III	1180.98	146.29	69.00	1.08
Genetic gain (%)	**37.85**	**12.99**	**32.40**	**0.83**

Biotechnological approaches

Efforts have been made by breeders in developing abiotic stress resistant varieties but are not sufficient enough. The above short hand of breeder can be overcome by the use of various biotechnological approaches as listed below:

1. QTL mapping

Quantitative trait loci (QTLs) are a region which contains or linked with a trait of interest which was governed by the genes (Miles and Wayne, 2008). Ruiz and Barbadilla (1995) stated that for a polygenic trait, to the most QTL govern 30 per cent for it contribution. This states that more the complex trait more will be the QTL number and lesser will be its contribution per se. The QTLs were identified by comparing genotypic and phenotypic data with a statistical package (R/QTL, MapQTL, QTL cartographer) to explain genetic variation due to particular set of gene (Lynch and Walsh, 1998). These traits are typically affected by the environment also. The following are the steps involved in QTL mapping:

Developing mapping population (F_2, RIL, NIL, BC)

↓

Genotyping by using molecular markers (SNP, SSR, CAPS, SCAR)

↓

Construction of linkage map from QTL analysis
(R/QTL, MapQTL, QTL cartographer)

↓

Phenotyping the mapping population (Field /glasshouse condition)

↓

Statistical analysis of genotypic and phenotypic data

Truco *et al.* (2000) identified several QTLs controlling shoot wilting under root chilling. Among them, the largest per cent of phenotypic effect (33%) was located on chromosome 9 in BC_1 mapping population. Later Goodstal *et al.* (2005) introgressed major QTL from *L. hirsutum* into a cultivated tomato using marker assisted backcross selection. Once the QTL was identified and mapped in particular geneotype or accession, it can be transferred to desirable background by repeated backcross or with any population improvement methods. The introgression can be accelerated by using marker.

2. Marker assisted selection

A prerequisite for marker assisted selection is that the gene governing trait of interest must be identified. As it is a indirect selection process where a trait of interest is selected, not based on the trait itself, but on a marker linked to it (Rosyara, 2006). MAS can be used since it is a non-distructive method of screening and useful screen or identify extremely complex trait and several traits can be screened simultaneously. Various approaches of MAS can be used with conventional breeding method to accelerate the success rate and to save the inputs.

Approaches

- Marker assisted backcross selection (MABC)
- Marker assisted early generation selection
- Marker assisted recurrent selection (MARS)

Marker assisted backcross selection (MABC)

Foreground selection (FG) :

A linked marker are used to select backcross progeny carrying the target gene

Background selection (BG):

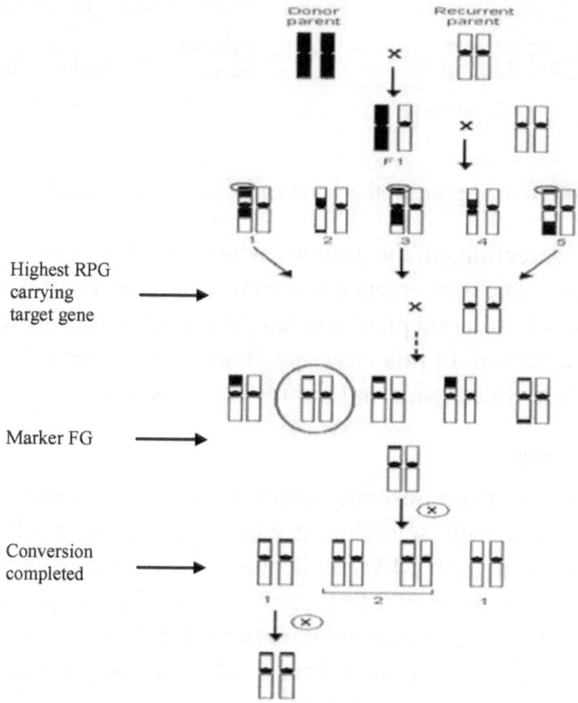

Random markers showing polymorphism with recurrent parent is use to select backcross progeny carrying recurrent parent genome

Marker assisted early generation selection

Instead of carrying all the plant progenies which are putative for containing target gene are selected on the basis of marker and such plants were proceed further in every selection cycle. This will save time and money of breeder and can carry desirable plants in every generation (Fig 3).

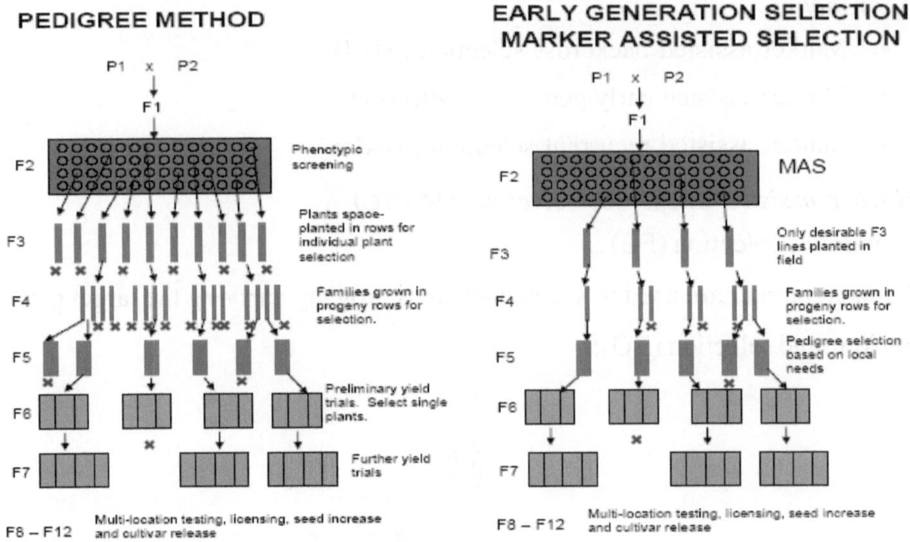

Fig 3. : Comparison of pedigree method and early generation selection with MAS.

While in recurrent selection, all the progeny which are assumed to have tolerant gene based on phenotypic performance is selected, then marker linked to tolerant gene are use to identify tolerant plant and later it was included in crossing cycle as like recurrent selection. In this case, the chance of getting tolerant plant or line is maximum and time, space and money can be saved.

3. Genomic selection

This is also known as genome-wide selection. This is effective because of availability of genome-wide selection at whole genome level (Heffner *et al.* 2009). This is different from MAS as it does not require marker associated with QTLs. It is a new form of MAS. This is a method to predict breeding values for individuals based on molecular markers distributed across the genome and compared with phenotypic data from a reference population (Kim *et al.* 2014). A diverse population is needed which is further grouped into Breeding population and training population. In former one, individuals are genotyped but not phenotyped while in later one, the entire population are genotyped with a large number of markers and phenotyped for important traits and GEBV is estimated and on that basis selection is made and later variety can be released (Fig 4).

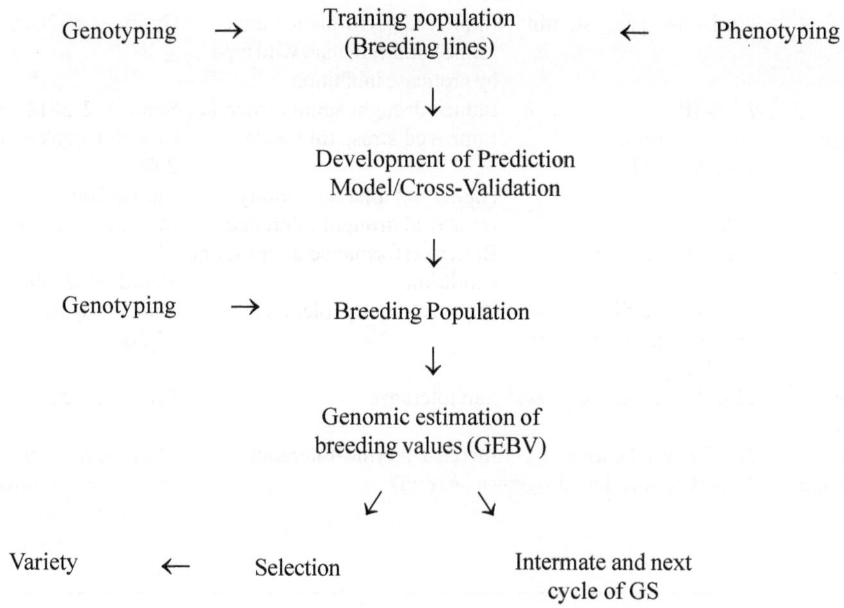

Fig. 4: Flow chart for genome-wide selection

4. Transgenic approaches

"Trans-gene" is an exogenous gene introduced in an organism and such organism is called transgenic which exhibit a new property which was not in it earlier. Various methods are there to developing transgenic among them most widely used is *Agrobacterium* mediated method. For introducing gene, a gene cassette is developed which contains promoter, coding region and terminator. The promoter enhances the expression by many times and terminator ends its expression profile so that downstream gene will not get affected.

Table 7 : List of abiotic stress tolerant transgenic in vegetable crops

Crop	Gene & Source	Trait improved	Reference
Tomato	*AtNHX1*(Arabidopsis)	Increased salt tolerance by over accumulation of Na^+ in vacuole	Apse *et al.* 1999
	tss1 (mutant of TSS1) (Tomato)	Increase uptake of K+	Borsani *et al.* 2001
	TPS1(Arabidopsis)	Osmo-protection	Garg *et al.* 2002
	CBF 1 (TF) (Arabidopsis)	Enhanced tolerance to chilling, water deficit and salt stress	Hsieh *et al.* 2002
	HsfA1		Mishra *et al.* 2002
	KatE(*Escherichia coli*)	Tolerance against photo-oxidative stress by drought	Mohamed *et al.* 2003

	Antisense prosystemin	Improved performance under saline condition than wild type by protinase inhibition	Orsini *et al.* 2010
	LIP9 (P)	Induce drought sainity tolerance	Shah *et al.* 2014
Potato	*rd29A* (P) and *DREB1A* (TF)	Improved stress tolerance	Celebi-Toprak *et al.* 2005
	Adi	Higher salt tolerance ability	Turhan 2005
	TPS-1	Improved drought tolerance	Stiller *et al.* 2008
	Catalase (bacterial)	Better performance under saline condition	Hamdi *et al.* 2009
	Δ^1-pyrroline-5-carboxylate synthetase (Arabidopsis)	Improve salinity tolerance	Hmida-Sayari *et al.* 2005
Water melon	*HAL1* (*Saccharomyces*)	Salt tolerance	Ellul *et al.* 2003
Carrot	*Hsp70* (Wild carrot)	Increase thermo-tolerance	Malik *et al.* 1999
Cabbage	*betA* (Choline dehydrogenase, *E. coli*)	Increase osmotic potential (300mM)	Bhattacharya *et al.* 2004

Note: P: Promoter, TF: Transcription factor

The main reason for the limited success of producing abiotic stress tolerant cultivars through genetic engineering is that in most cases only a single gene has been transformed, although it is now widely known that abiotic tolerance is multigenic in nature and a multitude of physiological, biochemical and molecular processes are involved in the mechanism of tolerance (Zhu, 2002). In majority of the studies, salt tolerance of transgenic lines has been appraised using only sodium chloride, although it is now well established that most soils contain significant amounts of salts other than sodium chloride (Buckman and Brady, 1967).

Conclusion and future projections

A focused research should be there by combining molecular, physiological and metabolic aspects of vegetable crops for bridging the present knowledge gaps. Conventional breeding can be supplemented with advance biotechnological approaches so that a target work can be achieved in short time with less input. A genome sequence of vegetable crops can also help in tackling and identifying the root cause for susceptibility and tolerance. Transgenic research opened new opportunities in vegetable crop improvement allowing gene transfer across living organisms. Many genes have been transferred but very few or nil have been commercialized. Multidisciplinary approach is needed for efficient crop improvement for abiotic stress tolerance.

Genome sequencing of important vegetable crops may open new secrets of their genus regarding biotic and abiotic stress tolerance. Molecular marker discovery and high throughput selection strategies could help in identifying exact genotype without wasting time and resources. Tolerance gene for abiotic stresses can give higher horizon of combined tolerance to different stresses and mining of gene regulation network and pathway for abiotic stresses could make it more clearly about its working principle which can later be incorporated and modified according to purpose. Functional genomics which includes various regulatory mechanisms along with functional processes could be applied to understand abiotic stress response.

References

Aamodt O.S. and Johnston W.H. (1936). Studies on drought resistance in spring wheat. *Can. J. Res.*, 14:122-152.

Anonymous (2011). Central Soil Salinity Research Institute, *Annual Report*: 2010-11. pp. 21-25.

Anonymous (2013). Reserve Bank of India. *Annual Report*. 2012- 13. pp. 13-26.

Apse M.P., Aharon G.S., Snedden W.A. and Blumwald E. (1999). Salt tolerance conferred by over expression of a vacuolar Na^+ /H^+ antiport in Arabidopsis. *Science*, 285: 1256–1258.

Ashraf M. and Akram N.A. 2009. Improving salinity tolerance of plants through conventional breeding and genetic engineering: an analytical comparison. *Biotechnol. Advances*, 27:744-752.

Ashraf M., Athar H.R., Harris P.J.C. and Kwon T.R. (2008). Some prospective strategies for improving crop salt tolerance. *Adv. Agron.*, 97: 45-110.

Baker R.J. (1988). Tests for cross-over genotype-environmental interactions. *Can. J. Plant Sci.*, 68: 405-410.

Bhattacharya R.C., Maheswari M., Dineshkumar V., Kirti P.B., Bhat S.R. and Chopra V.L. (2004). Transformation of *Brassica oleracea* var.*capitata* with bacterial beta gene enhances tolerance to salt stress. *Sci. Hortic*. 100:215–227.

Borsani O., Valpuesta V. and Botella M.A. (2001). Evidence for a role of salicylic acid in the oxidative damage generated by NaCl and osmotic stress in arabidopsis seedlings. *Plant Physiol*. 126:1024-1030

Buckman H.O. and Brady N.C. (1967). The Nature and Properties of Soils. New York: The MacMillan Company.

Carlson J.R., Ditterline R.L., Martin J.M., Sands D.C. and Lund R.E. (1983). Alfalfa seed germination in antiobiotic agar containing NaCl. *Crop Sci*. 23: 882-885.

Celebi-Toprak F., Behnam B., Serrano G., Kasuga M., Yamaguchi-Shinozaki K., Naka H., Watanabe J., Yamanaka S. and Watanabe K. (2005). Tolerance to salt stress of the transgenic tetrasomic tetraploid potato, *Solanum tuberosum* cv. Desiree appears to be induced by the DREB1A gene and rd29A promoter of Arabidopsis thaliana. *Breed Sci.*, 55: 311–319.

Chetelat R.T. (2015). Overcoming sterility and unilateral incompatibility of *Solanum lycopersicum* × *S. sitiens* hybrids. *Euphytica*, pp. 1-12

Chimungu J.G. and Lynch J.P. (2014). Root traits for improving nitrogen acquisition efficiency. In. Ricroch A., Chopra S. and Fleischer S.J. (eds.) Plant biotechnology: experience and future prospect. pp.181-192. Springer Cham Heidelberg New York Dordrecht London.

Cokus S.J., Feng S., Zhang X., Chen Z., Merriman B., Haudenschild C.D., Pradhan S., Nelson S.F., Pellegrini M. and Jacobsen S.E. (2008). Shotgun bisulphite sequencing of the Arabidopsis genome reveals DNA methylation patterning. *Nature*, 452: 215-219.

Cuartero J., Bolarin M.C., Asins M.J. and Moreno V. (2006). Increasing salt tolerance in the tomato. *J. Exp. Bot.*, 57:1045-1058.

Datta S.K. and Pardhan S.S. (1981). A screening method for salt tolerance of rice varieties at seelding stage. *Sci. Cult.*, 47: 444-446

Dauney M. (2008). Eggplant In: Prohens J, Nuez F and Carena M J (eds) Handbook of plant breeding Vegetable II: Fabaceae, liliaceae, Solanaceae and Umbelliferae. pp. 163-220. Springer Science and Business Media, New York, USA.

Doná A.A., Miranda G.V., DeLima R.O., Chaves L.G. and Gama E.E. (2012). Genetic parameters and predictive genetic gain in maize with modified recurrent selection method. *Chilean J. Agric. Res.*, 72: 33-39.

Donald C.M. (1968). The breeding of crop ideotypes. *Euphytica*, 17: 385-403.

Ellul P., Garcia-Sogo B., Pineda B., Rios G., Roig L.A. and Moreno V. (2003). The ploidy level of transgenic plants in Agrobacterium mediated transformation of tomato cotyledons (*Lycopersicon esculentum* Mill.) is genotype and procedure dependent. *Theor. Appl. Genet.*, 106: 231–238.

Falconer O.S. (1952). The problem of environment and selection. *Am. Nat.* 86: 293-298.

Foolad M.R. (2007). Genome mapping and molecular breeding of tomato. *Int. J. Plant Genomics*, 2007: 1-53.

Garg A.K., Kim J.K., Owens T.G., Ranwala A.P., Choi Y.D., Kochian L.V. and Wu R.J. (2002). Trehalose accumulation in rice plants confers high tolerance level to different abiotic stresses. *Proc. Natl. Acad. Sci.*, 99: 15898-15903.

González, C.M.C, Mukandama J.P., Fuentes J.L. and Alí M.M. (2005). Induction of Drought Tolerance in Tomato using ^{60}Co Gamma Ray Irradiation pp. 9-10. In: Shu Q and Lagoda P.J.L. (eds) Mutation breeding newsletter and review. *International Atomic Energy Agency* 1: 1-24.

Goodstal F.J., Kohler G.R., Randall L.B., Bloom A.J. and Clair D.A. (2005). A major QTL introgressed from wild *Lycopersicon hirsutum* confers chilling tolerance to cultivated tomato (*Lycopersicon esculentum*). *Theor. Appl. Genet.*, 111: 898-905.

Hall A.E. (2011). Breeding Cowpea for Future Climates. In: Yadav S.S, Redden R., Hatfield J.L., Lotze-Campen H. and Hall A.J.W. Crop Adaptation to Climate Change. Hoboken: John Wiley & Sons. pp. 340-355.

Hallauer A.R. (1985). Compendium of recurrent selection methods and their application. *Critic. Rev. Plant Sci.*, 3:1-34.

Hamdi M.M., Bettaieb T., Harbaoui Y., Mougou A.A., and Jardin P. (2009). Insight into the role of catalases in salt stress in potato (*Solanum tuberosum* L.). *J. Biotech. Agron. Soc. Environ.*, 13: 373–379.

Heffner E.L., Sorrells M.E. and Jannink J.L. (2009). Genomic selection for crop improvement. *Crop Sci.*, 49:1–12.

Hmida-Sayari A., Gargouri-Bouzid R., Bidani A., Jaoua L., Savoure A. and Jaoua S. (2005) Overexpression of Ä^1-pyrroline-5-carboxylate synthetase increases proline production and confers salt tolerance in transgenic potato plants. *Plant Sci.*, 169: 746–752.

Hsieh T.H., Lee J.T., Yang P.T., Chiu L.H., Charng Y.Y., Wang Y.C. and Chan M.T. (2002). Heterology expression of the Arabidopsis C-repeat/dehydration response element binding factor 1 gene confers elevated tolerance to chilling and oxidative stresses in transgenic tomato. *Plant Physiol.*, 129: 1086–1094.

Jana H.K. and Slinkard A.E. (1979). Screening for salt tolerance in lentils. LENS Newsl., 6: 25-27.

Kaufmann M.R. 1969. Effects of water potential on germination of lettuce, sunflower and citrus seeds. *Can. J. Bot.*, 47: 1761-1764.

Kim D.W., Agrawal G.K., Rakwal R., Ahmed S. and Rohila J.S. (2014). Genomic methods for improving abiotic stress tolerance in crops. In: Ricroch A., Chopra S. and Fleischer S.J. (eds) Plant Biotechnology: Expreience and Future prospects. pp. 35-44. Springer Cham Heidelberg New York Dordrecht London.

Kumar R., Solankey S.S. and Singh M. (2012). Breeding for drought tolerance in vegetables. *Veg Sci.*, 39: 1-15.

Latef A.A.H.A. and Ahmad P. (2015). Legumes and breeding under abiotic stress, in Legumes under Environmental Stress: Yield, Improvement and Adaptations. Azooz M.M. and Ahmad P. (eds), John Wiley & Sons, Ltd, Chichester, UK. pp. 1-20.

Lunin J., Gallatin M.H. and Batchelder. A.R. (1963). Saline irrigation of several vegetable crops at various growth stages-I effect on yields. *Agron. J.*, 55: 107-114.

Lynch M. and Walsh B. (1998). *Genetics and analysis of quantitative traits* (Vol. 1). Sunderland M.A. Sinauer. pp.319-351.

Malik M.K., Slovin J.P., Hwang C.H. and Zimmerman J.L. (1999). Modified expression of a carrot small heat shock protein gene, Hsp17. 7, results in increased or decreased thermotolerance. *Plant J.*, 20: 89-99.

Miles C. and Wayne M. (2008). Quantitative trait locus (QTL) analysis. Nature Education **1:** 208.

Mishra S.K., Tripp J., Winkelhaus S., Tschiersch B., Theres K., Nover L. and Scharf K.D. (2002). In the complex family of heat stress transcription factors, HsfA1 has a unique role as master regulator of thermotolerance in tomato. *Gene Dev.*, 16: 1555–1567

Mohamed E.A., Iwaki T., Munir I., Tamoi M., Shigeoka S. and Wadano A. (2003). Overexpression of bacterial catalase in tomato leaf chloroplasts enhances photo-oxidative stress tolerance. *Plant, Cell Environ.*, 26: 2037-2046.

Movahedi S., Tabatabaei B.E., Alizade H., Ghobadi C., Yamchi A. and Khaksar G. (2012). Constitutive expression of Arabidopsis DREB1Bin transgenic potato enhances drought and freezing tolerance. *Biol. Plant.*, 56: 37–42.

Mutava R.N., Prince S.J.K., Syed N.H., Song L., Valliyodan B., Chen W. and Nguyen H.T. (2015). Understanding abiotic stress tolerance mechanisms in soybean: A comparative evaluation of soybean response to drought and flooding stress. *Plant Physiol. Biochem.*, 6: 109-120.

Orsini F., Cascone P., Pascale S.D., Barbieri G., Corrado G., Rao R. and Maggio A. (2010). System independent salinity tolerance in tomato: Evidence of specific convergence of abiotic and biotic stress responses. *Physiol Plant.*, 138:10–21.

Pineda B., GarcíaAbellán J.O., Antón T., Pérez F., Moyano E., García Sogo B., Campos J.F., Angosto T., Morales B., Capel J. and Moreno V. (2012). Tomato: Genomic Approaches for Salt and Drought Stress Tolerance. Improving Crop Resistance to Abiotic Stress, Volume 1 & Volume 2. pp.1085-1120.

Rosenzweig C., Iglesias A., Yang X.B., Epstein P.R. and Chivian E. (2001). Climate change and extreame weather events. Implications for food production, plant diseases and pests. *Global Change and Human Health*. 2: 90-104.

Rosyara U.R. (2006). Requirement of robust molecular marker technology for plant breeding applications. *J. Plant Breed. Group*, 1: 67-72.

Roudier F., Ahmed I., Berard C., Sarazin A., Mary-Huard T., Cortijo S., Bouyer D., Caillieux E., Duvernois-Berthet E., Al-Shikhley L., Giraut L., Despres B., Drevensek S., Barneche F., Derozier S., Brunaud V., Aubourg S., Schnittger A., Bowler C., Martin-Magniette M.L., Robin S., Caboche M. and Colot V. (2011). Integrative epigenomic mapping defines four main chromatin states in Arabidopsis. *EMBO J.*, 30: 1928-1938.

Roy N.N. and Murty B.R. (1970). A selection procedure in wheat for stress environments. *Euphytica.*, 19: 509-521.

Ruiz A. and Barbadilla A. (1995). The contribution of quantitative trait loci and neutral marker loci to the genetic variances and covariances among quantitative traits in random mating populations. *Genetics*, 139(1): 445.

Sacher R.F. and Mulcahy D.L. (1981). Gametophytic selection for salt tolerance in tomato and related species. *Plant Physiol.,* 67: 19.

Shah S.H., Ali S., Jan S.A. and Ali G.M. (2014). Piercing and incubation method of in planta transformation producing stable transgenic plants by overexpressing DREB1A gene in tomato (*Solanum lycopersicum* Mill.). *Plant Cell, Tiss. Org.*, 120: 1139-57.

Sharma K.K. and Kumar S. (1985). Studies on salt resistance in wheat. Growth and development. *Curr. Sci.,* 54: 482-485.

Shrivastava P. and Kumar R. (2015). Soil Salinity: A serious environmental issue and plant growth promoting bacteria as one of the tools for its alleviation. *Saudi J. Biol. Sci.*, 22:123-31.

Stiller I., Dulai S., Kondrák M., Tarnai R., Szabó L., Toldi O. and Bánfalvi Z. (2008). Effects of drought on water content and photosynthetic parameters in potato plants expressing the trehalose-6-phosphate synthase gene of *Saccharomyces cerevisiae*. *Planta*, 227: 299–308.

Trachsel S., Kaeppler S.M., Brown K.M. and Lynch J..P (2010). Shovelomics: high throughput phenotyping of maize (*Zea mays* L.) root architecture in the field. *Plant and Soil,* 341:75–87.

Truco M.J., Randall L.B., Bloom A.J. and St. Clair D.A. (2000). Detection of QTLs associated with shoot wilting and root ammonium uptake under chilling temperatures in an interspecific backcross population from *Lycopersicon esculentum* × *L. hirsutum*. *Theor. Appl. Genet.*, 101:1082–1092

Turhan H. (2005). Salinity response of transgenic potato genotypes expressing the oxalate oxidase gene. *Turk. J. Agric.*, 29:187–195.

Zhu J.K. (2002). Salt and drought stress signal transduction in plants. *Annu. Rev. Plant Physiol. Plant Mol Biol.*, 53:247–273.

5

Exploiting Genetic Mechanisms for Heterosis Breeding in Brassica Vegetables

Saurabh Singh[1*] and Nidhi Sehgal[2]

[1]Division of Vegetable Science, ICAR-Indian Agricultural Research Institute New Delhi-110012, India

[2]Department of Vegetable Science, CCS Haryana Agricultural University Hisar, Haryana-125004, India

Abstract

Brassica vegetables belonging to family Brassicaceae are one of the most important groups of cool season vegetables. Initially, these crops were confined to Europe and other temperate regions. But, after the Second World War, their cultivation has spread to tropics and sub tropics as well. All the members of this group have originated primarily from the common ancestor wild cabbage (Brassica oleracea var. oleracea) and are fully cross-compatible with each other. In general, these are highly cross pollinated crops and have shown preponderance of non-additive gene action for most of the economic traits thereby suggesting the relevance of heterosis breeding. In the absence of appropriate genetic mechanism for the production of hybrid seed with ease at relatively lower cost, heterosis in a crop is of no practical value. Fortunately, the presence of natural genetic phenomenon of sporophytic self incompatibility and male sterility (particularly CMS) in brassica vegetables has proved instrumental in commercialization of hybrid seed production.

Keywords: *Heterosis, Self-incompatibility, Male Sterility, Cole Crops, Hybrid Breeding*

Introduction

Vegetables consist of many species and cultivars, and can be classified in many ways viz. botanical family and species, cultivar group, the edible part, climatic region, user ethnic origin, life cycle and nutrition. From the seed production point of view, they can be categorized into open-pollinated, F_1 hybrid and clonally propagated cultivars. The trend of F_1 hybrid seed usage in vegetable crops is increasing globally in terms of species, cultivars and volume of seed used. The popularity of F_1 hybrid cultivars is due to their vigor, uniformity, disease resistance, stress tolerance and good horticultural traits including earliness and long shelf-life expressed and thereby giving consistent stable high yield. From the breeders point of view, it is a fast and convenient way to combine desirable characters of a vegetable together, for example fruit size and color, plant type and disease resistance, high early and total yield and as a mean to control intellectual property rights through control and protection of the parental lines by the breeders.

The genus *Brassica* is the most economically important genus within the *Brassicaceae* family and belongs to the subtribe *Brassicinae*, one of the nine subtribes of the *Brassiceae* tribe that shares with other 18 tribes a wide gene pool, which over time has been utilized directly or indirectly to improve several crops (Branca and Cartea, 2011). Brassica vegetables are a family of 6 agriculturally important species consumed in high quantities throughout the world. The wild species of the *B. oleracea* group ($n = 9$) are found in small isolated areas and form very distinct phenotypes (Snogerup, 1980). The cole vegetables have been reported to be originated from common mediterranean centre of origin. In the western hemisphere, including Europe, the predominant species is *B. oleracea.* In Asia, *B. campestris* is the most cultivated species owing to the great importance of Chinese cabbage, although cabbage and cauliflower are extensively grown in Indian subcontinent. Turnip and turnip greens, which also belong to *B. campestris,* are cultivated worldwide but have much less economic importance. *B. juncea,* that includes the vegetable mustard, has some economic relevance in far eastern countries but is a minor crop at the world level. Vegetable *Brassicas* are cultivated in Europe since very ancient time from where they have spread to other parts of the world (Nieuwhof, 1969). The cytogenetic relationship of the main species of the *Brassica* genus was depicted in the U triangle (U, 1935; Fig. 1) which includes 3 monogenomic and 3 digenomic species (Table 1).

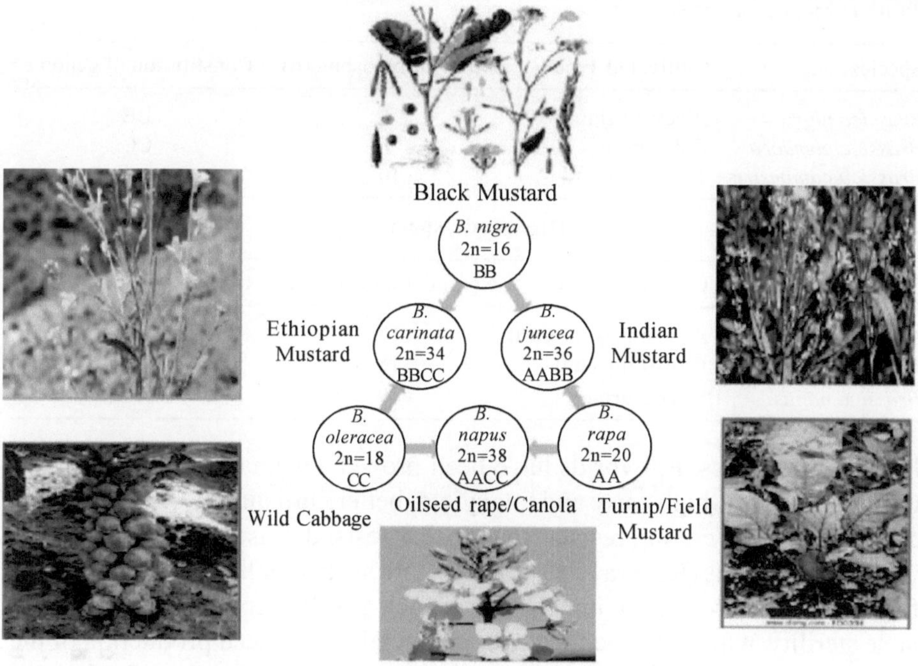

Fig. 1: Brassica U Triangle (U, 1935)

Cruciferous vegetables are a good source of many health promoting and potentially protective phytochemicals including folic acid, phenolics, carotenoids, selenium, glucosinolates and vitamin C. Incorporating these potent plant based compounds in a daily food is a safe, effective and inexpensive way to guard against many of today's most common and lethal cancers of which nearly 30-40% are directly linked to improper diet and related factors. *Brassica oleracea* is the main vegetable species of this family including kale, cabbage, cauliflower, broccoli, Brussels sprouts and knoll khol whereas vegetable forms of *Brassica rapa* include turnip, Chinese cabbage and pak choi (Cartea *et al.,* 2011). *Brassica* include many different morphotypes, which are well adapted to temperate climates, require quite simple cultivation techniques and produce abundant and nutritious food for man. Breeding strategy and targets are dependent on market trends. Successful breeders-anticipate changes in the market by developing new varieties that are ready to be released to the growers when their demand increases. It is, therefore, interesting to see how breeding is reacting to eventual changes in *Brassica* consumption and to evaluate the potential influence that the *Brassica* market and growing systems may have on the definition of breeding targets and priorities.

Table 1: Monogenomic species

Species	Cultivated type	No of chromosome(n)	Constitution of genome
Brassica nigra	Black mustard	8	BB
Brassica oleracea	Cole group	9	CC
Brassica campestris	Turnip group	10	AA

Digenomic species

Species	Cultivated type	No of chromosome(n)	Constitution of genome
Brassica carinata	Abyssinian mustard	17	BBCC
Brassica juncea	Brown mustard	18	AABB
Brassica napus	Swede group	19	AACC

In Cole vegetables, F_1 hybrids have been proved advantageous especially in uniform maturity, high early and total yield, better curd quality with respect to compactness and colour, resistance to insect pests, diseases and unfavourable weather conditions (Kucera *et al.,* 2006). There are two pollination mechanisms for hybrid seed production in cauliflower such as self-incompatibility (SI) and male sterility, which are used in the commercial hybrid seed production of the crop. Vegetable breeders have been very successful in utilizing them for decades in F_1 hybrid seed breeding.

Self-incompatibility (SI)

The term self-incompatibility was first coined by Stout, 1917. Self-incompatibility is the inability of a plant to set seed upon self pollination despite the fact that both male and female gametes are functional or inability of plant to set seed upon self pollination. Term incompatibility is used to describe the phenomenon, in which plants having functional gamete fail to set seed upon either self pollination or when crossed with some of their genetic relatives (Brewbaker, 1957). The self-incompatibility system is of two types, heteromorphic and homomorphic self-incompatibility. In the homomorphic type, there is no difference in flower morphology of same species, while in the heteromorphic system there is morphological difference in flowers of same species e.g. in primula (Kao and McCubbin, 1996). The homomorphic system is further categorized into gametophytic (GSI) and sporophytic (SSI) self-incompatibility, and basic difference between these two is presented in Table 2. The vegetables with highly developed self-incompatibility system are those in the family *Brassicaceae*. They include *Brassica oleracea* (Brussel's sprouts, cabbage, cauliflower, broccoli, kohlrabi and kale), *Brassica rapa* (Chinese cabbage, turnip and a range of Asian leafy Brassicas) and *Raphanus sativus*. The genetics of the self-incompatibility system in the cruciferous crops is so well developed that they consist of a series of genes (loci) and alleles. In the members of Brassica family including cole crops genetic mechanism of sporophytic self-incompatibility (SSI) exists and commercially exploited for hybrid breeding in cole vegetables (Kucera *et al.,* 2006; Shen *et al.,* 2008; Singh and Vidyasagar, 2012). SSI system has been confirmed in kale (Thompson, 1957), sprouting broccoli (Sampson, 1957), cabbage (Adamson, 1965) and cauliflower (Hoser-Krauze, 1979). However, its use in cabbage hybrid seed production had been demonstrated much earlier (Odland and Noll, 1950).

Table 2: Difference between GSI and SSI

	Gametophytic Self-incompatibility	Sporophytic Self-incompatibility
i.	Pollen-pistil interaction is genetically controlled by haploid genome of each of pollen grain and diploid genome of pistil tissue.	Pollen-pistil interaction is determined by pollen producing plant (sporophyte).
ii.	GSI is also known as haplo-diplo.	SSI is called as diplo-diplo.
iii.	Controlled by multiple alleles without dominance. Controlled by single locus except for grasses where two loci are involved.	Mostly controlled by single locus 'S' with multiple alleles showing dominance.
iv.	Stigma is smooth and wet.	Stigma is papillate (hairy) and dry.
v.	Site of pollen tube inhibition is in the style.	Pollen tube inhibition occurs on the stigmatic surface.
vi.	Reciprocal differences are not observed.	Reciprocal differences are observed.
vii.	GSI can be overcome by polyploidy.	It cannot be overcome by polyploidy.
viii.	Pollen tube is binucleate.	Pollen tube is trinucleate.
ix.	Homozygosity through normal pollination is not possible.	Homozygous individuals are possible.
x.	Incompatibility action is post-meiotic, i.e. depends upon the genetic status of pollen.	Incompatibility action is pre-meiotic, i.e. determined early by the genotype of pollen producing plant.
xi.	e.g. Fabaceae, solanaceae, rosaceae, liliaceae, papaveraceae	e.g. Brassicaceae, asteraceae, sterculaceae, convolvulaceae, caryophyllaceae

Molecular Mechanism of SI in Brassica

The SSI system in cole vegetables is controlled by single 'Mendelian S locus' (Bateman, 1955) with multiple alleles. Genomic analysis of S-locus in *Brassica* identified three physically linked loci *viz. SRK* (S-receptor kinase) (Stein, *et al.,* 1991), *SCR/SP11* (S-locus cysteine rich protein / S-locus protein 11) (Suzuki, *et al.,* 1999) and *SLG* (S-locus glycoprotein). *SRK* is the female determinant of self-incompatibility response in *Brassica,* a trans-membrane domain and an intracellular kinase domain (Stein et al. 1991). The *S* domain of *SRK* is highly similar to *SLG,* the first S-locus gene to be identified. Another gene reported at

S-locus by cloning and sequencing of S-locus region of *B. rapa*, encodes a protein expressed only in anthers, named as *SP11* (S-locus protein 11) or *SCR* (S-locus cysteine rich protein) and identified as male determinant of SI response in Brassica (Suzuki et al. 1999). *SCR/SP11*, *SLG* and *SRK* are tightly linked at S-locus and alleles of these genes are inherited by progeny as one set, called S haplotypes. Recognition specificity is achieved by interaction of stigmatic protein *SRK* (female determinant) with its ligand, the male determinant *SP11/SCR*. The interaction of *SCR* and *SRK* from the same *S* haplotype induces autophosphorylation of *S*-receptor kinase in an *S*-haplotype specific manner. When there is self-pollination, pollen ligand (*SCR* protein) is bound to maternal gene *SRK*, which causes the activation of receptor and elicit a signal transduction cascade that results in the pollen rejection and inhibition of pollen tube penetration into the epidermal cell wall of stigma (Nasrallah and Nasrallah 2014). Further, another protein *MLPK* (M-locus protein kinase) interact with *SRK*, and this complex induces signal transduction and result in self pollen rejection. In addition, the arm repeat containing protein 1 (*ARC1*), shows high specificity to *SRK*, and this *ARC1* is phosphorylated *by SRK-MLPK* complex, and phosphorylated *ARC1* degrades *Exo70A1*, thus inhibit the production of proteins for compatible reaction. Therefore, leads to self pollen rejection on selfing (Zhang *et al.*, 2011).

Steps to utilize SI phenomenon

a) Identification of self-incompatible plants in diverse population/genotypes; b) Development of homozygous self-incompatible lines; c) Identification of S-alleles in the homozygous self incompatible lines; d) Establishment of inter-allelic relationships among the S-alleles; e) Identifying the best combining lines; f) Maintenance of parental self-incompatible lines; g) Hybrid seed production on commercial scale.

Identification of self-incompatible plants is carried out by making different types of pollination *viz.* self pollination in freshly opened flowers (OP), compatible pollination with unrelated S-allele pollen (CP) and bud pollination to ascertain whether the plant in question is male and female fertile and also to get selfed seeds to raise its progeny. The plants are categorized self-incompatible/ compatible based on seed set in OP in relation to CP. For quicker (within 24-48 hours of pollination) identification of self-incompatible plants, fluorescence microscopy can be used (Vidyasagar and Chatterjee, 1984). Wallace (1979) suggested the use of seed set as well as pollen tube penetration (fluorescence microscopy) data. Homozygous self-incompatible lines are developed by making intra-progeny crossings/pollinations in freshly opened flowers in the progeny of a self- incompatible plant in full diallel by utilizing at least 7 progeny plants. The intra-progeny crossings/pollinations reveal heterozygous nature of the original self-incompatible plant. After grouping the progeny plants on the basis of compatible/incompatible reaction, we get at least two (one homozygous and one heterozygous) or all the three (both homozygous and one heterozygous) phenotypic groups. By selecting plants from the homozygous groups, homozygous lines could be developed in the next 2-3 generations. Information about the inter-allelic relationships among the S-alleles is important to ascertain their level of dominance and S-allele interactions. The S-alleles high in the dominance series are likely to give minimum selfs and sibs in hybrid seed production. S-allele interactions (in pollen and stigma) are worked out by using the data (seed set as well as pollen tube penetration) obtained from the pollinations carried out in freshly opened flowers reciprocally between a heterozygote (SxSy) and its two corresponding homozygotes (SxSx and SySy). Four types of S-allele interactions (Table 3) *viz.* type I (same S-allele dominant over the other in both pollen and stigma), type-II (one S-allele dominant over the other in pollen but co-dominant in stigma), type-III (one S-allele dominant over the other in stigma but co-dominant in pollen) and type-IV (both S-alleles co-dominant in pollen as well as stigma) have been reported. In order to have heterotic hybrids, it is essential to identify the best specific combining S-allele lines. This can be ascertained from the SCA (specific combining ability) studies and *per se* performance. Self-incompatible lines can also be combined with self-compatible lines but the hybrid seed quantities shall be lesser since the hybrid seed will be harvestable on the plants of self-incompatible lines only.

Table 3. Types of S-allele interactions in *Brassica* (Wallace, 1979)

Type	Pistil	Pollen
I	$S_1 > S_2$	$S_1 > S_2$
II	$S_1 = S_2$	$S_1 > S_2$
III	$S_1 > S_2$	$S_1 = S_2$
IV	$S_1 = S_2$	$S_1 = S_2$

The possibility of either co-dominant or dominant S-allele interaction occurring in the pollen and, independently of the same alternative co-dominant or dominant S-allele interactions occurring in the stigma, have been used to determine the incompatibility, compatibility and reciprocal-difference expectations for reciprocal crosses among the S-allele lines. These expectations are given for each of the sexual-organ x S-allele-interaction Types I, II, III and IV (Table 4).

Maintenance of Parental S-allele Lines

For the continuous hybrid seed production involving self-incompatible lines, the maintenance of parental S-allele lines is the basic requirement. Different approaches to maintain the parent self-incompatible inbred lines have been discussed in brief below:

I. Bud pollination (Shivanna, 1978).

II. Exposing the self incompatible plants to high concentration of CO_2 (3-5%) with relative humidity 100% in air tight chamber for a period of 8-24 hours (Palloix *et al.*, 1985). It requires high capital investment.

III. Spraying the solution of Sodium chloride (NaCl) at the full bloom stage at concentration of 3-5% followed by pollination in open flower stage results in a higher average seed set e.g. cabbage (Singh and Vidyasagar, 2012).

IV. Steel brush pollination (Roggen and Van Dijk, 1972) by mutilating the stigma during pollination with a brush of which the hair had been replaced by steel wire.

V. Electric aided pollination (Roggen and Van Dijk, 1973).

VI. Pollination at the end of flowering (Williams and Maier, 1977).

VII. Thermal aided pollination (Roggen and Van Dijk, 1976).

VIII. Tissue culture (Bhatia *et al.*, 2013)

Table 4: Expected incompatible and compatible interpretations and reciprocal difference interpretations from reciprocal intercrosses among two homozygous and one heterozygous SI lines for sexual organ x S-allele interaction types (Wallace, 1979)

Types of Interaction	S-allele interaction		Female genotype and phenotype	Male genotype and phenotype		
	Stigma	Pollen				
				SaSa	Sa < Sb	SbSb
I	D	D	SaSa	-	+	+
			Sa < Sb	+	-	-
			SbSb	+	-	-
II	Cod	D		SaSa	Sa < Sb	SbSb
			SaSa	-	+	+
			Sa = Sb	-	-	-
			SbSb	+	-	-
III	D	Cod		SaSa	Sa = Sb	SbSb
			SaSa	-	-	+
			Sa < Sb	+	-	-
			SbSb	+	-	-
IV	Cod	Cod		SaSa	Sa = Sb	SbSb
			SaSa	-	-	+
			Sa = Sb	-	-	-
			SbSb	+	-	-

Reciprocal crosses without reciprocal differences indicated by ⟷

Reciprocal crosses with reciprocal differences indicated by ←-------→

D: Dominance, Cod: Co- dominance; + (compatible), - (incompatible)

Commercial Hybrid Seed Production

Commercial hybrid seed production is done by way of developing single cross, three-way cross or double cross. In single cross, two self-incompatible but cross-compatible best combiners are planted in alternate rows in isolated plots. The hybrid seed is harvested on both the lines. In three-way cross, one single cross and a self-incompatible line are planted in alternate rows. Similarly in double cross, two single crosses are used. Pearson (1932) first suggested the utilization of genetic mechanism of self-incompatibility to facilitate hybrid seed production in *Brassica.* The commercial hybrid seed production was achieved in Japan in 1950, in the United States in 1954, and in Europe in 1966. Subsequently, Japanese companies produced 3- and 4- way hybrids by crossing self-incompatible but cross compatible inbreds (Wallace, 1979). The first 3-way cross hybrid using SI system in US was released in 1971 (Wallace, 1979). In USA, for hybrid seed production of cabbage, top cross is being used. For every 2 or 3 rows of a self incompatible line, one row of a good open pollinated (OP) cultivar as a pollen parent is provided. However, the hybrid seed is harvested from the self incompatible plants only.

Constraints in Exploiting SI Lines

- Continuous inbreeding in many *Brassica* crops may lead to complete loss of inbred lines. Therefore, it is more desirable to use tissue culture techniques for propagation of self-incompatible inbred lines.
- Pseudo-incompatibility may lead to self seed instead of hybrid seed e.g. cauliflower.
- Maintenance of S-allele lines is costly (hybrid seed will be expensive).
- Effect of environmental factors on the level of self-incompatibility (high temperature and relative humidity may weaken S-allele action).
- Synchronization of flowering among the SI lines is difficult e.g. cauliflower.
- Difficulties to produce homozygous inbred lines.

Male Sterility

In recent years, Brassica breeders are trying to use male sterility system instead of the standard incompatibility system. Male sterility is defined as the deviant condition in normally bisexual plants (monoecious as well as hermaphrodite) when no viable pollen is formed. Genic male sterility which is governed by recessive nuclear genes has been reported in cabbage (Rundfeldt, 1960), cauliflower (Nieuwhof, 1961) and other members of the cole group as well. In view of the problems associated with the maintenance of genic male sterile lines, its commercial use has been rather limited. In cauliflower, it has been utilized for hybrid production by maintaining the male sterile lines clonally. Some of the difficulties encountered were the reduction in nectary gland size and decreasing function of these glands. Plants with male-sterility gene were therefore unable to attract insect pollinators which are required for pollination.

Cytoplasmic Male Sterility

The occurrence of plants that lost their ability to produce viable pollen grains is frequent in hermaphroditic angiosperm species. As this is usually due to factors coded by the mitochondrial genome, this phenomenon is called cytoplasmic male sterility (CMS). The CMS is reported in over 150 plant species (Laser and Lersten, 1972) existing either as a spontaneous machinery (Hanson and Bentolila, 2004) or it can be created through experimental means like induced mutations, wide/inter-specific hybridization, protoplasmic fusion and genetic engineering (Yamagishi and Bhat, 2014). CMS system is currently commercially favoured for hybrid breeding in cole crops. Because the most reliable system of F_1 seed production is based on cytoplasmic male sterility (CMS), various types of CMS have been developed and adopted in practice to breed *Brassicaceae* oil seed and vegetable crops. CMS is a maternally inherited trait encoded by a gene located in the mitochondria. In CMS plants, pollen production is disturbed, whereas the function of the female organ is not generally affected. Expression of a mitochondrial CMS gene can be masked by nuclear fertility restorer (*Rf*) genes, allowing the plant to produce functional pollen. A combination of a nuclear genome that lacks *Rf* genes and a CMS-inducing mitochondrial genome leads to nuclear-cytoplasmic incompatibility and the CMS phenotype. CMS has been reported in a large number of plant species, and is very important both from practical breeding and from the fundamental genetic and evolutionary view points. CMS is expressed phenotypically when CMS-inducing mitochondrial genome is combined with a nuclear genome that lacks *Rf* genes. Such combinations can arise *de novo* by spontaneous mutation in the mitochondrial genome, outcrossing within a species, or artificial hybridization. The latter includes sexual interspecific or intergeneric crosses, and somatic cell fusions. Here, Brassicaceae CMS types are described according to their origins; (1) intraspecific variations, (2) alloplasmic origin by interspecific or intergeneric hybridizations, and (3) cell fusion (Yamagishi and Bhat, 2014).

Cytoplasmic male sterility has been advocated as an alternative to SI in cole crops to avoid problems associated with identification and maintenance of S-allele homozygote (Kucera *et al.*, 2006). However, in case of European summer cauliflower (Snowball/ Erfurt/Alpha), SI system is too weak to be utilized for hybrid seed production (Chatterjee and Mukherjee, 1965; Nieuwhof, 1963). The magnitude of self-incompatibility is negligible and stable SI lines are absent in snowball cauliflower (Sharma *et al.,* 1988). The commercial seed production of parental lines with strong SI is cumbersome, labor intensive and demands skilled personnel. CMS is the only viable alternative for hybrid seed production in cauliflower. The first report of cytoplasmic male sterility in Brassica is associated with Japanese radish (*Raphanus sativus* L.) or 'Daikon' (Ogura, 1968). Thereafter, many other CMS systems were described in Brassica by several workers. Among CMS types of *Brassicaceae*, the CMS discovered by Ogura (1968), which is now popularly referred to by his name, has been studied most extensively and is used worldwide in F_1 breeding of *B. napus*, *B. juncea*, *B. oleracea* and *R. sativus*.

Although Ogura system not well utilized in radish breeding in Japan, this CMS system was introduced into European radish. Some European radishes have an *Rf* gene (Bannerot *et al.,* 1977), whereas Ogura (1968) observed that Japanese cultivars have no such gene. European scientists introduced Ogura CMS into *B. napus* by intergeneric hybridization and repeated back-crossing (Bannerot *et al.,* 1974). The resultant alloplasmic lines of *B. napus* showed male sterility, but all of them had chlorotic leaves, yellowing at low temperatures (below 15°C) (Pelletier *et al.,* 1983). The introgression of radish Ogura cytoplasm into Brassica germplasm lead to the appearance of some undesirable characteristics associated, linked or even induced by the male sterile cytoplasm. These abnormalities were usually considered as a consequence of incongruity between the nucleus and cytoplasm (Pelletier *et al.,* 1983). Protoplast fusion was attempted to correct the problem of leaf yellowing. Few CMS lines have been developed at IARI, Regional Station, Katrain, (H.P.) through chloroplast substitution (Dey *et al.,* 2011).

CMS determining mitochondrial genes

The plant mitochondrial genome is highly complex and sensitive to frequent recombination, DNA rearrangements, incorporation of foreign DNA fragments, insertions/deletions and mutation (Chase, 2007; Bohra *et al.,* 2016). These characteristic features lead to generation of unusual chimeric open reading frames (*orfs*), which causes mitochondrial dysfunction and leads to cytoplasmic male sterility (Chase, 2007; Yamagishi and Bhat, 2014). About 14 mitochondrial genes determining CMS have been characterized as open reading frames (*orfs*) (Chase, 2007). Different *orfs* determining different origin of CMS have been characterized such as *pol* CMS contains a chimeric gene '*orf224*' which is co-transcribed with *atp6* gene, *ogu* CMS plants comprises '*orf138*' chimeric gene and co-transcribed with *atp8* gene, chimeric gene *orf222* determine *nap CMS*, similarly chimeric gene *orf288* co-transcribed with *atp6* gene determine *hau CMS* (Jing *et al.,* 2012; Yamagishi and Bhat, 2014). The association of a mitochondrial gene with CMS is inferred when the gene is expressed only in the male sterile plant but is suppressed in the fertility restored plant carrying a nuclear *Rf* gene.

Basic steps in the use of CMS

a) A-line (female parent) of desired genetic background (genetically Smsms); b) B-line (maintainer): This is an isogenic line in the genetic background as that of A-line and its function is in the maintenance of A-line (Figure 2) (genetically N-msms); c) C-line (male parent): This is a male parental line and also the best specific combiner with A-line. Genetically this could be N/S-Ms/- (fertility restoration) or N-msms since the economic product in cabbage and cauliflower is not the true seed; d) Maintenance of A, B and C lines; e) Hybrid seed production (Figure 3). The first increase of the seeds of the lines A, B and C must be in insect proof cages/structures followed by an increase in the open but not more than twice to obtain a high degree of purity in the stock seeds of the parental lines used for the production of F_1 hybrid seeds. The seeds of lines A and B can be produced in the same cage while the line C must be grown separately. Hybrid seed production on commercial scale is carried out in the open by providing recommended isolation distance of at least 1000 m. Usually for every 2 or 3 rows of A-line, one row of C-line is planted. The lines A and C must have perfect synchrony in flowering for good pollination and seed set. The hybrid seed is harvested from the plants of A-line only.

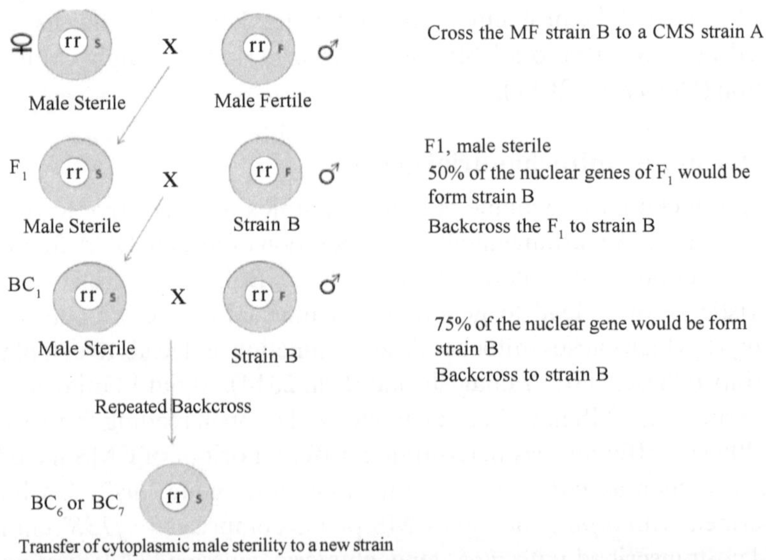

Fig 2. Maintenance of CMS lines. CMS is maintained by crossing it with pollinator strain used as recurrent parent in backcross programme as the nuclear genotype of pollinator is identical to new male sterile line, B line.

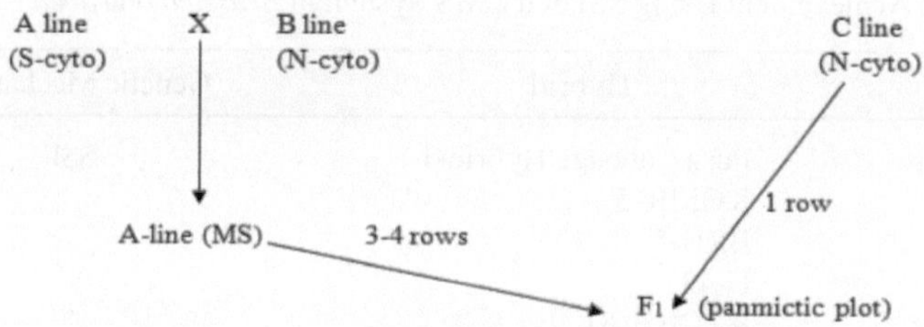

Fig. 3: Hybrid seed production using CMS system; cyto- cytoplasm

The main problem being faced is the lower quantity of hybrid seed on account of honey-bees preference for the pollen fertile C-line. This could be overcome to a certain extent with appropriate flower morphology and manipulating the ratio of plants of A and C lines, spacing and planting design. Inbreeding depression is another problem which results in low seed production of inbred parental lines and also that of single cross hybrids. In addition, three-and four way (double cross) crosses may prove economic.

Conclusion

In general, it can be concluded that both SI and CMS lines are suitable for hybrid breeding of cole vegetables. But as far as high seed yield is concerned, self-incompatibility appears to be more effective than cytoplasmic male sterility. On the contrary, CMS system provides much more reliable sterility than self-incompatibility. *Brassicaceae* crops display strong hybrid vigor, and have long been subject to F_1 hybrid breeding. Both the genetic phenomena have their own merits and demerits. The use of self-incompatible lines ensures the harvest of hybrid seed on all the plants in the hybrid seed production block. In CMS system, the hybrid seed will be harvested only from 67-75% of the plants in the hybrid seed production block. The latter phenomenon (CMS) has been considered better in avoiding selfs in the hybrid seed. However, in CMS lines also, some pollen production has been noticed especially at relatively higher temperatures nearing end of flowering stage thereby creating conditions for the possible selfs in the ultimate hybrid seed in this system as well. Because the most reliable system of F_1 seed production is based on cytoplasmic male sterility (CMS), various types of CMS have been developed and adopted in practice to breed *Brassicaceae* oil seed and vegetable crops. Several CMS genes have been identified by molecular genetic studies, including Ogura CMS from Japanese radish, which is most extensively studied and most widely used system.

Table 5: Breeding Achievement Using SSI and CMS System in *Brassica oleracea*

Crops	Hybrid	Genetic Mechanism
Cabbage	Pusa Cabbage Hybrid-1 KGMR-5 BRH-5 H-43 KTCBH-51 KTCBH-81	SSI
Cauliflower	Pusa Hybrid-2 Pusa Kartik Sankar Xiahua-6	SSI
Cauliflower	3 CMS lines developed at IARI-RS Katrain, Ogu1A, Ogu 2A and Ogu 3A Hybrids: KTH-51 KTH-52KTH-44 KTH-28 Pusa Snowball Hybrid-1	CMS
Cabbage	H-64KTCBH-84	CMS

References

Adamson, R.M. (1965) Self and cross-incompatibility in early round-headed cabbage. *Canadian Journal of Plant Science*, 45: 493-497.

Bannerot, H., Boulidard, L., Cauderon, Y. and Temp, J. (1974) Transfer of cytoplasmic male sterility from *Raphanus sativus* to *Brassica oleracea.* Eucarpia Meeting Cruciferae, Dundee, Scotland, pp. 52– 54.

Bannerot, H., Boulidard, L. and Cauderon, Y. (1977) Unexpected difficulties met with the radish cytoplasm. *Eucarpia Cruciferae Newsletter,* 2: 16.

Bateman, A.J. (1955) Self-incompatibility systems in angiosperms III *Cruciferae*. *Heredity*, 9: 52-68.

Bhatia, R., Parkash, C., Dey, S.S., Chandel, C. and Bhardwaj, V. (2013) *in vitro* propagation of a self-incompatible cabbage line 'Sel 5'. *Indian Journal of Horticulture*, 70: 364-368.

Bohra, A., Jha, U.C., Adhimoolam, P., Bisht, D. and Singh, N.P. (2016) Cytoplasmic male sterility (CMS) in hybrid breeding in field crops. *Plant Cell Reports*, 35: 967-993.

Branca, F. and Cartea, E. (2011) Brassica. In: C. Kole (ed.), Wild Crop Relatives: Genomic and Breeding Resources, Oilseeds, DOI: 10.1007/978-3-642-14871-2_2.

Brewbaker, J.L. (1957) Pollen cytology and incompatibility mechanisms in plants. *Journal of Heredity*, 48: 271-277.

Cartea, M.E., Francisco, M., Soengas, P. and Velasco, P. (2011) Phenolic compounds in Brassica vegetables. *Molecules*, 16: 251-280.

Chatterjee, S.S. and Mukherjee, S.K. (1965) Selection and maintenance of cauliflower (*B. oleracea* var. *botrytis* L.). *Indian Journal of Horticulture*, 22: 60–68.

Chase, C.D. (2007) Cytoplasmic male sterility: a window to the world of plant mitochondrial-nuclear interactions. *Trends in Genetics*, 23: 81-90.

Dey, S.S., Sharma, S.R., Parkash, C., Kumar, R. and Bhatia, R. (2011) Development and characterization of "Ogura" based improved CMS lines of cauliflower (*Brassica oleracea* var. *botrytis* L.). *Indian Journal of Genetics and Plant Breeding*, 71: 37–42.

Hanson, M.R. and Bentolila, S. (2004) Interactions of mitochondrial and nuclear genes that affect male gametophyte development. *Plant Cell*, 16: S154–S169.

Hoser-Krauze, J. (1979) Inheritance of self-incompatibility and the use of it in the production of F_1 hybrids of cauliflower. *Genetica Polonica,* 20: 341-367.

Jing, B., Heng, S., Tong, D., Wan, Z., Fu, T., Tu, J., Ma, C., Yi, B., Wen, J. and Shen, J. (2012) A male-sterility associated cytotoxic protein ORF288 in *Brassica juncea* causes aborted pollen development. *Journal of Experimental Botany*, 63: 1285-1295.

Kao, T.H. and McCubbin, A.G. (1996) How flowering plants discriminate between self and non-self pollen to prevent inbreeding. *Proceedings of National Academy of Sciences USA*, 93: 12059-12065.

Kucera, V., Chytilova, V., Vyvadilova, M. and Kalima, M. (2006) Hybrid breeding of cauliflower using self-incompatibility and cytoplasmic male sterility. *HortScience*, 33: 148-152.

Laser, K.D. and Lersten, N.R. (1972) Anatomy and cytology of microsporogenesis in cytoplasmic male sterile angiosperms. *Botanical Reviews*, 38: 425–454.

Nasrallah, J.B. and Nasrallah, M.E. (2014) *S*-locus receptor kinase signaling. *Biochemical Society Transactions*, 42: 313-319.

Nieuwhof, M. (1961) Male sterility in some cole crops. *Euphytica*, 10: 351-356.

Nieuwhof, M. (1963) Pollination and contamination of *Brassica oleracea* L. *Euphytica*, 12: 17–26.

Nieuwhof, M. (1969) *Cole Crops*. World Crop Books. London: Leonard Hill.

Odland, M.L. and Noll, C.J. (1950) The utilization of cross compatibility and self-incompatibility in the production of F_1 hybrid cabbage. *Proceedings of American Society for Horticultural Science*, 55: 391-402.

Ogura, H. (1968) Studies on the new male-sterility in Japanese radish, with special reference to the utilization of this sterility towards the practical raising of hybrid seeds. *Memories of the Faculty of Agriculture, Kagoshima University*, 6: 39–78.

Palloix, A., Herve, Y., Knox, R.B. and Dumas, C. (1985) Effect of CO_2 and relative humidity on self-incompatibility in cauliflower, *Brassica oleracea. Theoretical and Applied Genetics*, 70: 628-633.

Pelletier, G., Primard, G., Vedel, F., Chetrit, P., Remy, R., Rousselle, P. and Renard, M. (1983) Intergeneric cytoplasmic hybridization in Cruciferae by protoplast fusion. *Molecular Genetics and Genomics*, 191: 244–250.

Pearson, O.H. (1932) Breeeding plants of cabbage group. *California Agricultural Experiment Station Bulletin*, 532: 3-22.

Roggen, H.P.J.R. and Van Dijk, A.J. (1972) Breaking incompatibility of *Brassica oleracea* L. by steel brush-pollination. *Euphytica*, 21: 424-425.

Roggen, H.P.J.R. and Van Dijk, A.J. (1973) Electric aided and bud pollination: which method to use for self-seed production in cole crops (*Brassica oleraceae* L.)? *Euphytica*, 22: 260-263.

Roggen, H.P.J.R. and Van Dijk, A.J. (1976) "Thermally aided pollination": A new method of breaking self-incompatibility in *Brassica oleracea* L. *Euphytica*, 25: 643-646.

RUndfeldt, H. (1960) Untersuchungen Zur' Zllchtung des Koptkohls *(B. oleracea* L. var. *capitata). Z. Pjla: zenz*, 44 : 30·62.

Sampson, D.R. (1957) The genetics of self and cross-compatibility in *Brassica oleracea. Genetics,* 42: 253-263.

Sharma, S.R., Gill, H.S. and Kapoor, K.S. (1988) Heterosis and combining ability studies in late cauliflower (*Brassica oleracea* var. *botrytis* L.). *Vegetable Science*, 15: 55–63.

Shivanna, K.R. (1978). The pollen-stigma interaction: bud pollination in the cruciferae. *Acta Botanica Neerlandica*, 27: 107-119.

Shen, J.X., Wang, H.Z., Fu, T.D. and Tian, B.M. (2008) Cytoplasmic male sterility with self-incompatibility, a novel approach to utilizing Heterosis in rapeseed (*Brassica napus* L.). *Euphytica*, 162: 109-115.

Singh, S. and Vidyasagar. (2012) Effect of common salt (NaCl) sprays to overcome the self-incompatibility in the S-allele lines of *Brassica oleracea* var. *capitata* L. *SABRAO Journal of Breeding and Genetics*, 44: 339-348.

Snogerup, S. (1980) The wild forms of the *Brassica oleracea* group (2n = 18) and their possible relations to the cultivated ones. In: Tsunoda S, Hinata K, Gomez-Campo C (eds) Brassica crops and wild allies. Japan Scientific Societies Press, Tokyo, pp 121–132.

Stein, J.C., Howlett, B., Boyes, D.C., Nasrallah, M.E. and Nasrallah, J.B. (1991) Molecular cloning of a putative receptor protein kinase gene encoded at the self-incompatibility locus *Brassica oleracea. Proceedings of National Academy of Sciences USA*, 88: 8816-8820.

Stout, A.B. (1917) Fertility in *Cichorium intybus*. The sporadic occurrence of self-fertile plants among the progeny of self-sterile plants. *American Journal of Botany*, 4: 375-395.

Suzuki, G., Kai, N., Hirose, T., Fukui, K., Nishio, T., Takayama, S., Isogai, A., Watanabe, M. and Hinata, K. (1999) Genomic organization of the S locus: Identification and characterization of genes in *SLG/SRK* region of S_9 haplotype of *Brassica campestris* (syn. *rapa*). *Genetics*, 153: 391-400.

Thompson, K.F. (1957) Self-incompatibility in marrow stem kale *Brassica oleracea* var. *acephala*. I. Demonstration of sporophytic system. *Journal of Genetics*, 55: 45-60.

Wallace, D.H. (1979) Interactions of S-alleles in sporophytically controlled self-incompatibility of *Brassica. Theoretical and Applied Genetics,* 54: 193-201.

Wallace, D.H. (1979) Procedures for identifying S-allele genotypes of *Brassica*. *Theoretical and Applied Genetics*, 54: 249-265.

Williams, R.R. and Maier, M. (1977) Pseudo-compatibility after self pollination of apple cox's Orange pippin. *Journal of Horticultural Science*, 52: 475-483.

Yamagishi, H. and Bhat, S.R. (2014) Cytoplasmic male sterility in Brassicaceae crops. *Breeding Science*, 64: 38–47.

Zhang, X., Yin, D., Zhu, W., Ma, C. and Fu, T. (2011) Progress on characterization of self-incompatibility in *Brassica napus* L. *Euphytica*, 182: 147-155.

Section 02
Improved Propagation Techniques

6

Recent Propagation Techniques of Horticultural Crops

***Km. Lalita*[1] *and Abha Kumari*[2]**

[1]*Department of Applied Plant Science (Horticulture)*
Babasaheb Bhimrao Ambedkar University, Lucknow-226025, India
[2]*Department of Horticulture (Fruit & Fruit Technology)*
Bihar Agricultural University, Sabour, Bhagalpur-813210, India

Introduction

India is a rich center of plant diversity due to varied climatic conditions as well as soil qualities which provide a better platform for the diverse plant groups. In India around 45000 plant species are spread over the area (Mayers, 1988). These plant species multiply and produces the new plant. The multiplication of plants is termed as propagation and the art of propagation is as old as civilization of human being. The history of civilization began with the propagation and growing of plants. Plant propagation was in practice in pre-historic period. However, the propagation process has got commercial popularity only recently. The success in plant propagation depends on a thorough knowledge of the growth of plants, choice of suitable method and technical skill. In nature, plants multiply naturally to perpetuate themselves. Now-a-days, propagation practices have been totally modernized and became advanced from field to *in vitro*. Variation in plant propagation can be observed in the nature like some plants are producing many seeds for future generation with their unique type of dispersal mechanism. Among a large diversity of the plants, many plants does not produces seeds and are equally distributed by propagating itself following another mode of propagation like by bud, bulb, rhizome, tuber, corm, stem cutting etc. Some plants are also capable to regenerate by both the methods like by their Seeds and Stem cutting. Starting from seeds or scion stick of 15-20 cm length, the plants are multiplied even using cell, tissue and embryo. Tissue culture or *in vitro* culture is known as the micropropagation. Today plant tissue culture applications encompass much more than clonal propagation. The range of routine technologies has expanded to include somatic embryogenesis, somatic hybridization, virus elimination as well as the application of bioreactors to mass propagation. Perhaps the greatest value of these tissue culture technologies lies not so much in their application to mass clonal propagation but rather in their role underpinning developments and applications in plant improvement, molecular biology and bioprocessing, as well as being a basic research tool. In modern horticulture trade, plant propagation is quite profitable, employment generative and creative business.

Types of Propagation

The details of the methods are presented below-

Sexual propagation

Sexual propagation is the raising of the plants by means of seeds which are formed due to the fusion of male and female gametes within the ovule of a flower. Meiosis division occurs in course of fusion and the chromosome numbers, as in parents, are reduced to half which after fertilization becomes normal. The plants thus produced may or may not be similar to their parents and the propagated plant may also be different from each other (Singh, J., 2002).

Merits of sexual propagation

- Sexual propagation is easier and cheaper than asexual propagation.
- Seed propagation is the only means of creating diversity, chance seedling and essential for genetic improvement of plant.
- Seed produced plants are long lived, produce and show greater tolerance to adverse soil, climate and disease endemics as compared to vegetatively propagated plants.
- Hybrids are developed easily through seeds.
- Seed propagation makes feasible to propagate plants like papaya and coconut in which asexual means of propagation is not common.
- Some plants like citrus, mango and jamun have polyembryony
- Polyembryonic seeds produce seedlings from integuments tissues and nuclear cells of embryo. Such seedlings being free from effect of fertilization are identical to mother plant. Thus, there is no need of practising asexual propagation in polyembryonic seeds.
- For asexual propagation rootstocks are raised by seeds.

Demerits of sexual propagation

- Most of the fruit plants are heterozygous in nature. Multiplication of these plants by sexual means results into segregation of characters and the resultant plants are never true-to-type.
- Seedlings have long phase of juvenility and the first crop is obtained very late.
- There is no uniformity in growth, yield and quality of fruits.
- The seedling pants are long and gigantic in stature. Harvesting and spraying becomes difficult.
- Viruses like Psorosis in citrus and viruses in peach and cherry are transmitted through seeds.
- It is not possible to perpetuate the exact characters of any superior selection.
- In seedlings, it is not possible to take advantage of the modifying influence of rootstock on the scion.

Development of seeds: Successful pollination and fertilization are the prerequisites to produce a viable seed. Like fruit, the seed also exhibits sigmoid and sometimes double sigmoid pattern of growth. Various changes occur during the course of seed development and the sequences of development may vary from species to species keeping the basic theme unaltered.

Seed production: Seed production of any variety is a very specialized technique. A breeder's seed is an offspring of nucleus seed. The breeder seed is further multiplied into the foundation and certified seeds. For the production of the superior quality of the seeds, the seed plot should be located in a place with abundant sunshine. The recommended agronomic package and practices in respect of seed treatment, seed rate, fertilizer dose, irrigation, weeding and plant protection measures must be followed to obtain a healthy crop. The proper isolation distance from plot to plot should be maintained depending on the mode of pollination to avoid natural cross pollination. Delay is harvesting may reduce the quality of seed. Proper care should be taken during mechanical manipulation of seed to prevent from damage and varietal admixture. Storage of seed is an important aspect, which needs proper attention, to maintain viability.

Germination of seed- Seed germination refers to the resumption of metabolic activity and growth of an embryo, resulting in the rupture of seed coat and emergence of plantlets (Sadhu, 1989). The prerequisites for germination process are:

a. Seed must be viable
b. It must be non dormant and the embryo is quiscent
c. The seed must be placed in conductive atmospheric conditions which vary from species to species (Ching, 1972; Jann and Amen, 1977).

Asexual/Vegetative propagation

Mitosis is the basic process of normal vegetative growth, regeneration and wound healing which make possible the vegetative propagation techniques like cuttage, graftage, layerage, separation and divisions. These methods of propagation are important because they permit a large-scale multiplication of a genetically identical individual plant.

Merits of asexual/vegetative propagation

- Asexually propagated plants are true to type to their mother plants.
- The propagated plants are smaller in stature and hence harvesting and spraying becomes easy.
- Using asexual method, it is possible to convert a non productive local variety into productive improved type variety.
- Repairing of damaged portion of plant is possible in asexual propagation. Bridge grafting is suitable method for such purpose.
- Asexually propagated plants possess short juvenile phase and flowers and fruits early in the age.
- Some of the fruit crops do not produce viable seeds (*e.g.* banana, pineapple, fig etc.) and multiplication of them is only possible through asexual means.

Demerits of asexual/vegetative propagation

- Asexually propagated plants have short life.
- The vegetatively propagated plants particularly budded / grafted are short-lived. No new varieties can be evolved by vegetative means of propagation.
- Asexual propagation plants restrict vegetational diversity.
- In citrus, tristeza virus (quick decline) is transmitted through bud wood. Thus, sometimes asexual propagation disseminates diseases.

Methods of vegetative propagation

1. Apomictic seed
2. Specialized vegetative structure- Bulbs, corms, rhizomes, tubers etc.
3. On the own root system- Cutting, layering.

4. On root system of other plants- Budding, grafting.
5. Micropropagation- meristem culture, embryo culture, callus culture etc.

1. Apomictic seed: Development of embryo without fertilization are called apomixis and the seedling produced by this means is known as apomictic seedling.

Types of apomixis

a) Recurrent apomixis

b) Non- recurrent apomixis

c) Adventitious or nucellar embryony

d) Vegetative apomixes

2. Specialized vegetative structure: Propagation by specialized organs can be done easily. Various modified plant organs like bulb, corm, tuber, rhizome etc, which are primarily storage organ for food, are used as planting materials. Two different methods are employed for propagation by the modified plant parts:

a) Separation- naturally detachable structures- *e.g.* bulb, corm.

b) Division- cut into sections- *e.g.* rhizome; tuber.

3. On the own root system

a) Cutting: Separation of a portion from mother plant and planting it in a media suitably so that it may constitute a new plant successfully is termed as cutting.

Types of cutting: Cutting is of three types-

1) Stem cutting
2) Root cutting
3) Leaf cutting

1) **Stem cutting**: When a portion of stem is taken for propagation, called stem cutting. Nanda and Kochhar (1987) recorded that stem cutting as a popular plant propagation method. It is of four types-

   i. **Hard wood cutting-** In this method, during November-February one year old and mature shoots are selected for the purpose of propagation. The length of cutting is kept to 25-45 cm. Grape, fig, pomegranate, morus and roses are propagated through hard wood cutting.

   ii. **Semi-hard wood cutting-** In this type of cutting, 4 to 9 months old shoot, 7 to 20 cm. length treated with IBA @ 5000 ppm is used for raising new plants during rainy season. Mango, guava, jackfruit, lemon, aonla etc.

iii. **Soft wood cutting:** Shoots of 2-3 months age and 10-15 cm. length are selected for soft wood cuttings. Apple, peach, guava and many ornamental plants can be propagated under mist chamber using soft wood cuttings.

iv. **Herbaceous cuttings:** Many ornamental plants are planted through herbaceous cuttings. Shoots of 1-2 months age are selected for cuttings. *Alternanthera, Coleus, Pilea, Iresine* etc. are propagated by herbaceous cutting.

2) **Root cutting:** For propagation, roots of 2 to 3cm. thickness and 10 to 15 cm. length are selected. For temperate fruit, it is practised in December while for subtropical fruits, it is practised during rainy season. The prepared cutting is placed in sphagnum grass or moist sand medium for callusing. Fruit plants like Apple, pear, guava and bael and forest species are suitable for root cutting.

3) **Leaf cutting:** Mostly succulent plants having thick and leathery leaves are propagated through leaf cutting. Leaf is separated from mother plant and planted in suitable medium for rooting. Begonia, sansevieria, crassula etc. are propagated through leaf cutting.

b) Layering : Layering involves the regeneration of adventitious roots from a shoot when it is still attached to the mother plant. Girdling and etiolation positively affect the rooting in layering. The effect of etiolation was first demonstrated by Sachs as early as 1964 (Sachs, 1965) and was subsequently corroborated by several workers (Delargy, 1979). The regulatory mechanism of this method is still obscure although it is possible that light interferes or causes destruction of root promoting action of endogenous auxin (Hartmann and Kester, 2009). Layering is commonly practised with great success in many fruit crops like, guava, litchi, citrus etc.

Types of Layering- There are six types of layering-

i. Simple layering

ii. Serpentine/Compound layering

iii. Mound layering/Stooling

iv. Trench layering

v. Tip layering

vi. Air layering

i. **Simple layering:** Low flexible, generally one year old shoot is used for layering. The shoot is bent downward in the soil using peg or nail and tied with the help of rope to make it stayed in the position. In about 4 to 5 months, roots emerge out. It is then separated from the mother plant. Guava and hazelnut can be propagated through simple layering.

ii. **Serpentine/Compound layering:** The branches of the plants covered with the soil at their nodes throughout its length by alternate exposing of internodal length of shoot is termed as compound layering. eg. Muscadine grape.

iii. **Mound layering/Stooling:** Before the commencement of the new growth, the plant is pruned at 2.5 cm above ground level, subsequently a large number of shoots emerge out, shoots are girdled at the base and covered with the soil to a height of about 15 to 20 cm, rooting occurs in about 3 to 4 months, then shoot is separated from mother plant and planted in the nursery.eg. Guava, apple, pear, quince, currants etc.

iv. **Trench layering:** In this method, horizontal shallow trench is prepared and the branch of the plant bent downward in it. when the shoots develops, the base is covered with 5 to 10 cm layer of soil., the rooted layers are separated from mother plants and planted in nursery. eg. apple, pear and cherry.

v. **Tip layering:** In this method, the top of the plant is layered by making a hole 2 to 5 cm below the soil. Tip roots out and produce shoots then it is separated from the mother plant along with the old shoot to handle the new rooted tip.eg. Blackberries and raspberries.

vi. **Air layering:** It is also called chinese layering, Pot layering, Marcottage or Gootee. About 5 to 7 cm away from the base of selected shoot, a girdle of 2.5 to 3.0 cm size, by removing the bark, is prepared. Girdled portion is scrapped using gunny bag, covered using moist sphagnum moss grass and wrappped using transparent polyethylene tape. Rooting occurs in about 2-3 months. The layered plants are planted in nursery under partial shade.eg. Litchi, lime, jackfruit etc.

4. On root system of other plants

a) Budding : The technique of combining the bud and rootstock in such a way that they may unite successfully. The process of budding is done when there is active sap flow in plant. According to Lee (2000), conventional budding has been successfully performed on one to four month old seedlings in Malaysia.

However, Yow and Lim (1994) reported that conventional budding applied to three-month-old seedling rootstocks takes atleast six months to produce field plantable budded seedlings

Types of budding- There are seven types of budding-

1. Shield budding
2. Patch budding
3. Chip budding
4. Ring budding
5. Modified ring budding
6. Forkert budding
7. Modified forkert budding

    1. **Shield budding:** When the bud is inserted by making vertical incision on rootstock, it is termed as shield budding while the incision is made in T- shape, it is termed as T-budding.eg. Apple, pear, rose, peaches, apricot etc.
    2. **Patch budding:** In this method, square or rectangular shape bud is taken out from scion shoot. Similar size incision is made on rootstock. The bud is placed on it. Polyethylene tape is wrapped on bud leaving the sprouting portion exposed. eg. Jackfruit, aonla, mango, jamun etc.
    3. **Chip budding:** Chip budding is practised during the period when there is lack of sap flow. The bud is taken out from scion shoot along with wood. Similar size incision is made on the rootstock, both are wrapped with polyethylene tape. eg. Apple, pear, grapes.
    4. **Ring budding:** In this method ring shape bark containing a bud and 2.5 to 3.0 cm length is taken out from scion shoot and it is slipped downward by rotating it. No wrapping is required. eg. Ber, peach and morus.
    5. **Modified ring budding:** In this method, instead of taking out complete circular ring of bark containing bud, the bud wood is taken out from scion shoot by giving a vertical slit in bud wood.eg. Guava, ber, walnut etc.
    6. **Forkert budding:** In forkert method, the piece of bark is not removed and it is kept hang to the stock plant. A rectangular piece of bark containing the bud is taken from scion plant which fits well in the cut surface of stock and flap is drawn over and wrapped using polyethylene tape. eg. Mango, jackfruit, cashewnut etc.

7. **Modified forkert budding-** This method is similar to the forkert budding but the vertical flap of the bark is covered to base portion of the bud only.

b) Grafting : Grafting is a technique in which scion stick and rootstock is connected in such a way that they may unite and grow successfully as single component plant.

Types of grafting: Grafting is divided into following types-

1. **Inarching:** In this method, the rootstock is grown in pot, the rootstock is brought close to the scion, both are of pencil thickness, shallow cut of 2.5-4.0cm length is prepared on scion. Similar matching cut is also given on rootstock. Both rootstock and scion brought close to each other and wrapped, when union takes place, the upper portion of rootstock is cut. The scion shoot is detached from mother plant. eg. Mango, jackfruit, custard apple etc.
2. **Whip/Splice grafting:** In this method, one year old rootstock is used and scion of matching thickness is used. Both rootstock and scion are united and tied using polyethylene tape until union becomes completed. eg. Walnut, Apple, pear etc.
3. **Tongue grafting:** In this method a slanting cut similar to whip grafting is prepared both on rootstock and scion, another cut is also made just opposite to the previous. Scion is inserted into the rootstock and interlocked, wrapped by polythene tape. eg. Apple, pear etc.
4. **Side grafting/ Veneer grafting:** In this method a terminal shoot of pencil thickness is selected and a V-shaped incision is made both on rootstock and scion, both are fastened together using 300 guage polyethylene tape. eg. Mango.
5. **Soft wood grafting:** In this method one year old plant having pencil thickness is selected for rootstock and 3-5 months of age 10-15 cm length scion stick is selected. The rootstock is beheaded at 2.5-4.0 cm length, a vertical slit is made on rootstock and similar cut is made on scion also. It is inserted and wrapped using polyethylene tape. eg. Mango.
6. **Bridge grafting:** In this technique, the damaged portion of rootstock is scratched, in healthy portion of rootstock incision is made on top and bottom portion of the stock. The scion portion of suitable length is inserted into incision. It is fixed using nail and seal using wax. eg. Apple, pear, cherry etc.

7. **Epicotyl/Stone grafting:** In this method 7-10 days age seedling is used for grafting. The seedling is deheaded at a height of 10 cm from ground level. A vertical slit of 2.5 to 4 cm length is given on deheaded portion of rootstock. Scion shoot of 2-3 months age having pencil thickness is used, scion is grafted on rootstock and wrapped using polyethylene tape. The grafted plants are then maintained in other beds. eg. Mango.
8. **Cleft grafting:** Studies in Malaysia showed that cleft grafting and nurse-seed grafting produced an acceptably high level of end success, varying from 70% to 80% (Ramadasan and Ahmad 1986). In this method, on the stub to be grafted, a downward vertical split of 5 to 8 cm length is prepared. The vertical incision is made on two sides of stock, scion shoots pencil thickness is used, matching incision is prepared on lower portion of scion. The scion is inserted in vertical slits of rootstock and wrapped. eg. Mango, hazelnut, pecannut, grape etc.
9. **Wedge grafting:** In wedge grafting, V-wedge shape incision of about 5 cm length is prepared on the stub of the plant. Shallow and downward matching incision is prepared on lower portion of scion, scion is inserted in rootstock and waxed.

5. Micropropagation: Micropropagation is the practice of rapidly multiplying stock plant material to produce a large number of progeny plants, using modern plant tissue culture methods. Single cells, plant cells without cell walls (protoplasts), pieces of leaves, or (less commonly) roots can often be used to generate a new plant on culture media given the required nutrients and plant hormones (Vidyasagar, 2006). The German Plant Physiologist Haberlandt (1902) first described the biological principles of tissue and organ culture. The basic protocols were well established by the 1960s and a whole research field and industry grew based on the ubiquitous MS medium (Murashige and Skoog, 1962) and the numerous modifications that have followed.

At present, tissue culture finds extensive application in agriculture and horticulture in several countries. Though some achievements have been made but the commercial utilization of the techniques of tissue culture is still not upto desire.

Merits of Micropropagation

1. Micropropagation or tissue culture helps in rapid multiplication of true-to-type plants throughout the year.
2. Micropropagation works on the principle of totipotency. A new plant can be developed from a miniature plant part.

3. Many more plants can be produced at a time in small space with uniform growth and productivity.
4. Micropropagation facilitates long term storage of clonal materials and long distance transport.
5. Micropropagation produced the plants free from viruses and diseases.
6. Micropropagation is very useful for those plants that don't breed true from seeds, seeds are not viable or available easily like banana.

Demerits of Micropropagation

1. Very high cost involved in setting up and maintenance of laboratory
2. Tissue culture techniques require skill and manpower.
3. A tiny mistake or slight infection may damage the entire lot of plants.
4. Generally, it is observed that the plant developed from tissue culture technique fail to survive under environmental condition unless hardening is done.

Methods of Micropropagation: There are mainly five methods of micropropagation.

1. Meristem culture
2. Callus culture
3. Cell culture
4. Embryo culture
5. Protoplast culture

1. Meristem culture- In meristem culture the meristem and a few subtending leaf primordial are placed into a suitable growing media. An elongated rooted plantlet is produced after some weeks, and is transferred to the soil when it has attained a considerable height. This technique can be successfully utilised for rapid multiplication of various herbaceous fruit plants. The success in achieving such directed morphogenic events are largely determined by the cultured tissue itself. Several explant-related factors appear to influence the organogenic potential of the cultured tissue (Benson, 2000).

2. Callus culture- When a living plant tissue is placed in an artificial growing medium, with other conditions favourable, callus is formed. The callus growth and its organogenesis or embryogenesis can be classified into three different stages-

i. Rapid production of callus after placing the explant in culture medium.

ii. The callus is transferred to other medium containing growth regulators for the induction of adventitious organs.

iii. The new plantlet is then exposed gradually to the environmental condition.

3. Cell culture- A cell suspension culture refers to cells dispersed and growing in an aerated liquid culture medium (Street, 1977; Thorpe, 1981). A piece of soft tissue is placed in a liquid medium and shaken vigrously so as to obtain a suspension of cells. Suezawa *et al.* (1988) reported development of cytokinin induced adventitious buds in kiwi fruit in a suspension culture subcultured for about a week.

4. Embryo culture- In embryo culture, the embryo is excised and placed into a culture medium with proper nutrient in aseptic condition. To obtain a quick and optimum growth of the embryo, the culture medium is changed 2 to 3 times. When the embryo has grown into a plantlet, it is transferred to soil. It is particularly important for the production of interspecific and intergeneric hybrids and to overcome the embryo abortion.

5. Protoplast culture-The Protoplasts are first cultured in liquid medium at 25° to 28°C with a light intensity of 100-500 lux or in dark and after undergoing substantial cell division, they are transferred into solid medium congenial for morphogenesis. Many horticultural crop responses well to protoplast culture.

References

Benson EE (2000). *in vitro* plant recalcitrance: An introduction. *in vitro*: Cell. Dev. Biol. Plant.

Chadha, K.L.(2005), Handbook of Horticulture, Indian Couuncil of Agricultural Research, Directorate of Information and Publications on Agriculture.

Chattopadhyay, T.K. (2012) A Textbook on Pomology (Fundamentals) Vol-1, Kalyani Publishers, New Delhi, India.

Ching, T.M. (1972), Metabolism of germinating seeds, in T.T. Kozlowski, (ed). Seed Biology, Vol.2. New York, Academic Press.

Delargy J. A. and Wright C.E. (1979). Root formation in cuttings of apple in relation to auxin application and to etiolation. New Phytol 82.

Hartmann H T, Kester D E, Fred T D, Robert L, Geneve R L. (2009). Plant Propagation Principles and Practices, 7 edition, Prentice Hall, New Delhi.

Jann, R.C. and Amen, R.D. (1977). What is germination? In the Physiology and biochemistry of seed dormancy and germination, A.A. Khan, (ed) Amsterdam, North Holland Publishing Co.

Lee, M.T. (2000). Recent experiences in field use of cocoa clones for large scale commercial planting in Malaysia: pros and cons, Proceedings of the technical meeting: state of knowledge on mass production of genetically improved propagules of cocoa, October 1998, Ilhéus, Bahia, Brazil.

Mayers, N. (1988). Threatened biotas: Hotspots in tropical forests. Environmentalist 8.

Murashige T, Skoog, F. (1962). A revised medium for rapid growth and bioassays with tobacco tissue cultures. *Physiol. Plant*, 15:473-97, 1962.

Nanda, K. K. and Kochhar, V. K. (1987). Vegetative propagation of Plants. Pribciples and Practices, India Book Trust.

Ramadasan, K., Ahmad, M.Y. (1986). Recent advances in vegetative propagation techniques of Theobroma cacao L., under Malaysian conditions Proceedings of the International conference on cocoa and coconuts, 15–17 October 1984, Kuala Lumpur.

Sachs, R.M. (1965) Stem elongation. Annu Rev Plant Physiol.

Sadhu, R.N. and Kaul, V., 1989, Seedcoat dormancy in Robinia pseudo- acacia, Indian Forester, 121.

Singh, J. (2002). Basic Horticulture, Kalyani publishers, New Delhi, India.

Vidyasagar K (2006). National Conference on Plant Biotechnology, Lady Doak College, Madurai.

Yow, S.T.K, Lim, D.H.K. (1994). Green-patch budding in very young cocoa rootstock and side-grafting of mature trees, Cocoa Growers' Bulletin, 47.

7

Micropropagation in Woody Perennials

Suhasini Chikkalaki* and Ashwini Ganur

Department of Fruit Science, College of Horticulture
University of Horticultural Sciences, Udyanagiri, Bagalkot-587 104
Karnataka, India

Abstract

A perennial plant or simply perennial is a plant that lives for more than two years. The term is often used to differentiate a plant from shorter-lived annuals and biennials. The term is also widely used to distinguish plants with little or no woody growth from trees and shrubs, which are also technically perennials. in vitro technologies are continuously expanding in the field of biology. Plant tissue culture has become a general title for a very broad subject. Commercial use of micropropagation is still limited, because of its relatively high production cost resulting mainly from high labor costs, low growth rate in vitro, and poor survival rate of the plantlets during acclimatization. Micropropagation of woody perennials is essential due their importance in production of secondary metabolites, which are playing a vital role now a days. Stages of micropropagation i.e mother plant selection, establishment of the aseptic culture, multiplication, root regeneration and transfer to the natural environment, which are systematically reproduced in many protocols which will aids in micropropagation studies of many woody perennials. Thus the review of present study attempts to provide a comprehensive account on micropropagation of woody perennials complementing conventional programmes for improvement.

Keywords: *Micropropagation, Woody perennial, in vitro, Aseptic culture and Acclimatization.*

Introduction

A perennial plant or simply perennial is a plant that lives for more than two years. The term is often used to differentiate a plant from shorter-lived annuals and biennials. The term is also widely used to distinguish plants with little or no woody growth from trees and shrubs, which are also technically perennials. The use of tissue culture techniques for woody perennials has increased considerably over the past few decades. Although its commercial use as a method for multiplying plants still fairly new, tissue culture propagation (Micropropagation) has already had a significant impact on the way people think about and handle plants. As a technique, micropropagation represents a direct, practical extension of scientific methodology devised over 30 years ago to study fundamental aspects of plant physiology. The ever increasing demand for many forest products and the progressive deterioration of natural forests means that the forest industry cannot continue to rely on the exploitation of natural forests (Jain, 1997; Tzfira *et al.*, 1998). To meet the increasing demand for forest products while more forest land is needed for non-timber uses, the replacement of natural populations of trees with cultivated varieties is absolutely crucial. At present, short rotation intensive culture of hardwoods appears to augment this need very well. Research in conventional propagation and breeding has made tree growers more aware of their inability to vegetatively mass propagate many tree species. At the same time, failure to produce a number of desirable inter-specific and inter-generic hybrids due to incompatibility and the long life span of trees has raised interest in the techniques of genetic engineering.

In vitro technologies are continuously expanding in the field of biology. Plant tissue culture has become a general title for a very broad subject. While in the beginning it was possible to culture plant cells either as established organs, such as roots or as disorganized masses, it is now possible to culture plant cell in a variety of ways, individually (as single cells in microculture systems), collectively (as calluses or suspensions, on petri-dishes, in Erlenmeyer flasks or in large-scale fermentors), or as organized units as shoots, roots, ovules, flowers, fruits etc. However, in the recent past tissue culture technology has been increasingly used in highly applied contexts. Successes in a number of areas have been achieved. Micropropagation deals with the propagation of plants, *in vitro*, has many advantages over conventional vegetative propagation. Its application in horticulture, agriculture and forestry is currently expanding world-wide. The goal of micropropagation is to mass-produce genetically identical, physiologically uniform, developmentally normal and pathogen-free plantlets, which can be acclimatized in a reduced time period and at a lower cost. Development of both automated environmental control systems and improved *in vitro* culture systems are essential for a significant reduction in production cost. However, commercial use of micropropagation is still limited, because of its relatively high production cost resulting mainly from high labor costs, low growth rate *in vitro*, and poor survival rate of the plantlets during acclimatization. Altman and Loberant have elegantly reviewed principles and practices of micropropagation.

Problems Associated with Tissue Culture in Woody Perrenials

Browning of media

High phenolic exudation owing to activation of explant and media browning has made it hard for workers dealing with woody plants. Mostly when mature plant parts such as nodal explants, shoot tips, leaf segments are used for culture establishment of many woody plants. Browning is major hindrance in micropropagation of many woody perennials due to exudation of phenolic compounds. Thus the methods commonly used to overcome the harmful effect of browning are; the use of adsorbing agents such as activated charcoal or polyvinylpyrrolidone (PVP), addition of antioxidants to the culture medium or soaking the explants in antioxidant solution, transfer of explants to fresh medium at frequent intervals or a combination of approaches with varying degree of success (Mahishi *et al.* 1991; Naik *et al.* 1999; Al-Wasel 1999; Murkute *et al.* 2003, 2004; Singh and Khawale 2006; El-Agamy *et al.*, 2009).

in vitro recalcitrance

The inability of plant cells, tissues and organs to respond to tissue culture is often referred to as '*in vitro* recalcitrance' (Benson 2000). Morphogenic responsiveness of explants cultured *in vitro*, is largely a net positive result of interplay between the innate physiological status, precisely the endogenous levels of phytohormones of the donor tissue and the influence of the exogeneously added plant growth regulators (PGRs) in the culture medium. Reports (Moriguchi *et al.* 1987; Omura *et al.* 1987; Naik and Chand 2003) showed that, woody perennials, if not a recalcitrant one, at least for *in vitro* regeneration through organogenesis. Of the several factors responsible for *in vitro* recalcitrance of a plant species, production of ethylene in in vitro culture is a key factor. Can be over come by ethylene inhibitors namely silver nitrate or aminoethoxyvinylglycine (AVG) (Naik and Chand 2003; Kanwar *et al.*, 2010).

Contamination (Bacterial/Fungal)

Other major bottleneck in micropropagation of woody perennials is contamination may caused by either bacteria or fungi. Severe or devastating diseases of many plants have made them more prone to contamination in tissue culture. One best example is bacterial blight of pomegranate. It can be overcome by selection of genuine explants from healthy mother plants. As a source of disease may be mother plant if bacteria/fungi is endogenous in nature. Antibiotics are used in some difficult cases.

Shoot tip necrosis

Shoot tip necrosis occurs in some woody perennial tissue cultures when actively growing shoot tips develop tip die-back. Shoot tip necrosis occurs in some woody perennial tissue cultures when actively growing shoot tips develop tip die-back. This condition is usually caused by a calcium deficiency in the medium. It can be rectified by addition of sufficient amount of calcium in media.

Habituation

Habituation is when a culture continues to develop in the absence of auxin or cytokinin. For example, shoot cultures habituated for cytokinin would continue to produce new shoots on a cytokinin free medium.

Stages in Micropropagation of Woody Perennials

According to Professor Murashige, all methods of plant micropropagation involve three basic types of manipulations, designated mainly as Stage I, Stage II, and Stage III. According to Debergh and Maene (1981) that such preparative procedures should be called Stage 0. A fourth stage (IV), at which plants are transferred to the external environment, is now also commonly recognised.

Stage 0 : Mother plant selection and preparation: Before micropropagation commences, careful attention should be given to the selection of stock plants. They must be typical of the variety or species, and free from any symptoms of disease. It may be advantageous to treat the chosen plant (or parts of it) in some way to make *in vitro* culture successful. Steps to reduce the contamination level of explants (Vol. 2) were considered sufficiently important by Debergh and Maene (1981) to constitute a separate essential stage in a commercial micropropagation programme. It seems appropriate to include all procedures adopted in plant selection and pre-treatment within 'Stage 0'. The recognised numbering of Murashige's stages is then unaltered.

Stage 1 : Establishment of the aseptic culture: An "explant" (part of a stock plant) is cleaned, disinfected, and placed on a tissue-culture medium. The objective of Stage I is to obtain a living and growing plant tissue free from microbial contamination. Surprising as it may seem, this goal is usually the most difficult thing to achieve in micropropagation. It is carried out in presence of initiation medium.

Stage 2 : Multiplication: sometimes coincides with Stage I, especially when shoot multiplication is used. The aim of Stage II is the rapid increase in shoots or other structures that ultimately can give rise to plants. Often in shoot multiplication, explants respond almost immediately to the high cytokinin concentration of the medium by proliferating new shoots. In these cases, Stage II shoot multiplication has the potential of producing one million shoots in a year, starting from a single growing tip.

Stage 3: Root regeneration: Involves all the manipulations required for establishment of tissue-culture derived plants in soil. If, for instance, shoot multiplication is used for Stage II, then Stage III technology consists of a rooting treatment that produces plantlets and then a gradual process of acclimation (hardening) of these plantlets to the lower humidity and increased light intensity of the greenhouse or outdoor environment. Depending on the tenderness of the plantlets obtained from tissue culture, hardening may last two to eight weeks.

Stage 4: Transfer to the natural environment: Although not given a special numerical stage by Murashige, the methods whereby plantlets are transferred from the *in vitro* to the *ex vitro* external environment are extremely important. In practice, plantlets are removed from their Stage III containers, and if they have been grown on agar medium, the gel is carefully washed from the roots. The application of an anti-transpirant film to the leaves has been recommended at this stage, but in practice, seems to be seldom used. Plantlets are then transplanted into an adequate rooting medium (such as a peat:sand compost) and kept for several days in high humidity and reduced light intensity.

Stage	**Methodology Involved**
	↓
Stagge 0	Selection of mother plant and its maintenance ↓
Stage I	Initiation and establishment of culture ↓
Stage II	Multiplication of shoots or rapid somatic embryo formation ↓
Stage III	*in vitro* germination of somatic embryos and/or rooting of shoots ↓
Stage IV	Transfer of plantlets to sterilized soil for hardening under greenhouse environment

Process of Micropropagation

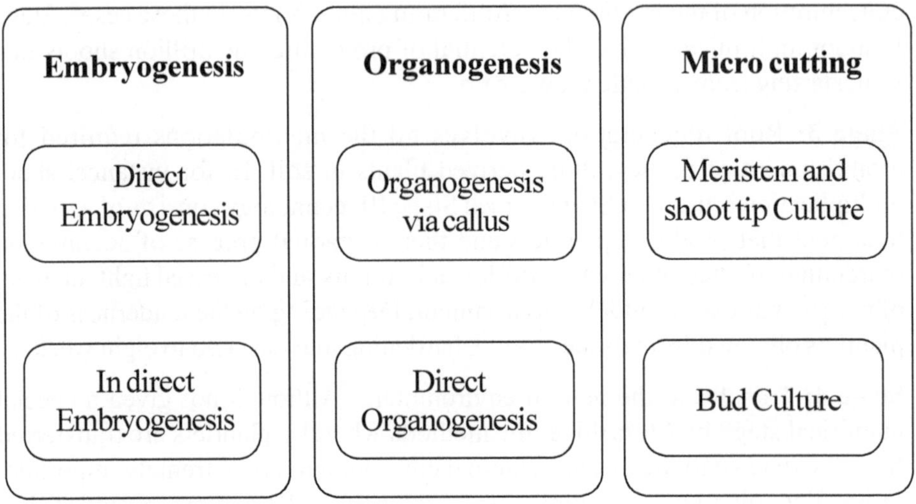

Fundamental Abilities of Tissue Culture Plants

- **Totipotency**: It is ability of a single cell to regenerate into whole plant
- **Dedifferentiation**: It is ability matured cells to revert back to its meristamatic stage to form unorganized mass of cells
- **Competency**: Ability of cells to regenerate into the organs like shoots and roots
- **Plasticity**: Ability of plantlets to alter their meyabolism with respect to the new environment.

Advantages of Micropropagation of Woody Perrenials Over Conventional Method

Characters	Micropropagation	Conventional method
Vegetative growth	Fast and more vigorous	Slow and less vigorous
Stem / stump	Thick stump so it can bear maximum number of fruits with weight. So easy to take more yield on tree and less damage of stump breaking	Thin stump, so limitation in taking more yields. Chance of breaking the stump due to more fruits
Leaves	More number of leaves with maximum branches	Less leaves with minimum branches
Pest/disease infestation	less	more
Flower formation	Early	Late

Applications of Micropropagation of Woody Perrenials

Stress studies

Trees, by their perennial nature and immobility, are constantly exposed to a variety of biotic and abiotic stress factors. Stress induces a variety of physiological and developmental changes that can be long-term or short-term in nature. Whereas some of these changes are quantitative, i.e. increase or decrease in cell size or the cellular content of one or more metabolites, others are qualitative, e.g. the presence or absence of certain metabolites or proteins.

Secondary metabolite production

Plants are not only the primary source for food and fiber, but also a source of pharmaceutical and industrial products such as drugs, dyes, scents, etc. However only a small number of the compounds sold commercially are produced in cell cultures, primarily because yields are low (Alfermann and Petersen, 1995). The topic of plant tissue cultures for the production of industrial biochemicals has been thoroughly reviewed (Staba, 1980; Fujita, 1990; Bajaj, 1995, Cosson, 1997; Yeoman and Yeoman, 1997). The major advantages of producing secondary metabolites in cell tissue cultures are: (1) Year round availability of cultures; (2) Only specific cellslorgans can be grown in vitro; (3) secretory products can be easily purified; (4) production of a single compound can be enhancedthrough the use of chemical elicitors, biotransformation or by transgene expression; and (5) the production of recombinant biopharmaceuticals such as mammalian hormones, enzymes, vaccines and monoclonal antibodies.

Host – pathogen interaction studies

Tissue cultures have been used successfully for studying a variety of host pathogen interactions, including the mechanisms of disease resistance and apoptosis, at the cellular and molecular level (Able *et al*., 1998; Furukawa and Kunihiko, 1998; Hehers and Sonnewald, 1998; Ragazzi *et al*., 1998). Most of the work published thus far pertains to herbaceous plants. The progress in this area of research helps to demonstrate the usefulness of these studies to woody plants in the near future. *in vivo* studies to analyze host pathogen interactions and resistance mechanisms are confounded by site and microclimate variability, physiological age of the tissues at the inoculation site, and infections by other pathogens besides the one under study, along with abiotic stresses. The synthesis, regulation, and activities of pathogen induced compounds such as phytoalexins, phytotoxins, polyphenols, and pathogenesis related proteins (PW's) can be studied in response to exposure of cultured explants and tissues to pathogens. However, one must bear in mind the possibility that the behavior of the host or the pathogen can prove to beartifactual under *in-vitro* conditions.

Conclusion and Prospective

Establishment of reliable protocols for regeneration of plants from tissue explants and callus cultures has been instrumental in the success of genetic engineering techniques with crop plants. The phenolic exudations (browning), *in vitro* recalcitrance, shoot tip necrosis, habituation and contamination in *in vitro* multiplication of woody plants have made them restricted to regenerate vigorously. Although tremendous studies have been made in this regard with woody species, including forest trees, the step of large scale multiplication, especially via somatic embryogenesis from morphogenetically stable cultures, is still a major impediment for widespread use of transformation techniques with trees. The difficulty of regeneration from protoplasts, and the fact that only juvenile material can be used for tissue culture in most tree species, further compounds the situation. A better understanding of the biochemical and molecular basis of habituation would help prevent this phenomenon and achieve continued regeneration of shoots or somatic embryos in cell cultures. There is a need for the development of methods for recognizing cell populations with high potency for regeneration within mixed cultures. The relatively long lag periods needed for evaluation of improved phenotypes necessitates the development of biochemical and molecular markers that can aid in an early selection of the desired genotypes.

References

Able, AJ., Guest, D.I., Sutherland, M.W. (1998). Use of new tetrazolium-based assay to study the production of superoxide radicals by tobacco cell cultures challenged with avirulent zoospores of *Phytophthora paracitica* var nicotiana. *Plant Physiol.* 117:491-499.

Alfermann, A.W. and Petersen, M. (1995). Natural products formation by plant cell biotechnology. *Plant Cell Tiss. Organ. Cult.* 43:199-205.

Al-Wasel, A.S.A. (1999). *in vitro* clonal propagation of "Al-Belehi" pomegranate (*Punica granatum* L.). *J King Saud Univ* 11:3–14.

Amin, M.N., Islam, M.N. and Azad, M.A.K. (1999). Regeneration of plantlets *in vitro* from the seedling explants of pomegranate (*Punica granatum*). *Plant Tissue Cult.* 9:53–61.

Bajaj, Y. P. S. (ed.). (1995a). *Biotechnology in Agriculture and Forestry 30. Somatic Embryogenesis and Synthetic Seed.* Springer-Verlag, Berlin. 472p.

Benson, E.E. (2000). *in vitro* recalcitrance: an introduction. *in vitro Cell Dev Biol Plant* 36: 141–148.

Cosson, L. (1997). Plant biotechnology contribution in study and production of natural plant substances. *Acta Botanrca Gallica.* 144:463-468.

El-Agamy, S.Z., Mostafa, R.A.A., Shaaban, M.M. and El-Mahdy, M.T. (2009). *In vitro* propagation of Manfalouty and Nab El-gamal pomegranate cultivars. *Res J Agric Biol Sci* 5:1169–1175.

Fujita, Y. (1990). The production of Industrial compounds. In: *Plant Tissue Culture: Applicatrons and Limitations,* 259-175 pp. (Ed. S.S. Bhojwani). Elsevier Science Publishers, New York.

Furukawa, T., and Kunihiko, S. (1998). Increased production of IAA by *Rhizoctonia solani* is induced by culture filtrate from rice suspension cultures. *Plant Cell Physiol.* 39:43-48.

Herbers, K and Sonnewald, U. (1998). Altered gene expression brought about inter and intracellulary formed hexoses and its possible implications for plant-pathogen interactions. *J. Plant Res.* 1:323-328.

Jain, S.M. (1997). Biotechnology of industrially important tree species in developing countries. In: *Plant Biotechnology and Plant Genetic Resources for Sustainability and Productivity* (eds. K.N. Watanabe and E. Pehu), 227-238 pp. Academic Press, New York.

Kanwar, K., Joseph, J. and Deepika, R. (2010). Comparison of in vitro regeneration pathways in *Punica granatum* L. *Plant Cell Tissue Organ Cult.,* 100:199–207.

Kumar, A. and Roy, S. (eds) Plant Biotechnology and its applications in tissue culture. IK International Pvt. Ltd., New Delhi, pp 107–113.

Mahishi, D.M., Muralikrishna, A., Shivashankar, G., Kulkarni, R.S. (1991). Shoot tip culture method for rapid clonal propagation of pomegranate (*Punica granatum* L.). In: Prakash J, Pierik RLM (eds) Horticulture-new technologies and applications: proceedings of the international seminar on new frontiers in horticulture. Kluwer Academic Publishers, Dordrecht, pp 215–217.

Moriguchi, T., Omura, T., Matsuta, N. and Kozaki, I. (1987). *in vitro* adventitious shoot formation from anthers of pomegranate. *Hortic Sci.,* 22:947–948.

Murkute, A.A., Patil, S. and Mayakumari. (2003). Exudation and browning in tissue culture of pomegranate. *Agric Sci Digest* 23:29–31.

Murkute, A.A., Patil, S. and Singh, S.K. (2004). *in vitro* regeneration in pomegranate cv. Ganesh from mature trees. *Indian J Hortic* 61:206–208.

Naik, S.K., Chand, P.K. (2003). Silver nitrate and aminoethoxyvinylglycine promote *in vitro* adventitious shoot regeneration of pomegranate (*Punica granatum* L.). *J Plant Physiol* 160:423–430.

Naik, S.K., Pattnaik, S. and Chand, P.K. (1999). *in vitro* propagation of pomegranate (Punica granatum L. cv. Ganesh) through axillary shoot proliferation from nodal segments of mature tree. *Sci Hortic* 79:175–183.

Omura, M., Matsuta, T., Moriguchi, T. and Kozaki, I. (1987). Adventitious shoot and plantlet formation from cultured pomegranate leaf explants. *Hortic Sci.* 22:133–134.

Ragazzi, *A,* Moricca, S., and Dellavalle, I. (1998). Growth of *Fusarium eumartii* on callus tissue of oak. *Pjlanzenkrankheiten-Pjlanzenschutz*. 105:279-282.

Singh, S. K. and Khawale, R.N. (2006). Plantlet regeneration from nodal segments of pomegranate (*Punica granatum*). *Plant Biotechnology and its Applications in Tissue Culture*. 12: 107-113.

Staba, E.J. (1980). Plant Tissue Culture as a source of Biochemicals. CRC Press, Boca Raton, FL.285 p.

Yeoman, M.M. and Yeoman, C.L. (1997). Manipulating secondary metabolism in cultured plant cells. *New Phytologist,* 134:553-569.

8

Micropropagation and Its Utilization in Spice Crops

Suddhasuchi Das, Sarthak, Bhattacharya, T. Adhikary A. Bandyopadhyay, and D.K. Ghosh

Department of Spices and Plantation Crops, Faculty of Horticulture Bidhan Chandra Krishi Viswavidyalaya, Mohanpur-741252, West Bengal

Abstract

Plant research often involves growing new plants in a controlled environment. These may be plants that we have genetically altered in some way or may be plants of which we need many copies all exactly alike. These things can be accomplished through tissue culture of small tissue pieces from the plant of interest. The latest list of ISO recognizes 109 spices, but only a few are important commercially such as black pepper, cardamom, ginger, turmeric, large cardamom, cumin, coriander, fennel, fenugreek, garlic, saffron and celery. They are extensively used in flavouring food and confectionery, perfumery and soaps and also as colouring materials. Many of these groups of crops have medicinal properties and are used in many Ayurvedic and Unani medicines. Inspite of the rich diversity and importance, the productivity of spices is low in India. Spices are an important group of economic plants with large genetic diversity in the Indian subcontinent and need to be conserved through in-vitro methods. Conservation of the germplasm using traditional method of field maintenance is problematic, normally labour intensive and also face risk of germplasm loss due to pest/pathogen attack or natural disasters. To minimize the problems of their conservation and increasing interest has been shown to use plant tissue culture. In vitro conservation of germplasm has been reported in various crop species including the spices black pepper, cardamom, ginger, turmeric, large cardamom.

***Keywords**: Tissue culture, spices, black pepper, turmeric, in vitro conservation*

Introduction

Plant research often involves growing new plants in a controlled environment. These may be plants that we have genetically altered in some way or may be plants of which we need many copies all exactly alike. These things can be accomplished through tissue culture of small tissue pieces from the plant of interest. These small pieces may come from a single mother plant or they may be the result of genetic transformation of single plant cells which are then encouraged to grow and to ultimately develop into a whole plant. Tissue culture techniques are often used for commercial production of plants as well as for plant research.

Spices are an important group of economic plants with large genetic diversity in the Indian subcontinent and need to be conserved through *in- vitro* methods. Conservation of the germplasm using traditional method of field maintenance is problematic, normally labour intensive and also face risk of germplasm loss due to pest/pathogen attack or natural disasters. To minimize the problems of their conservation and increasing interest has been shown to use plant tissue culture. (Bhowal *et al.*, 2011). *In vitro* conservation of germplasm has been reported in various crop species including the spices ginger and turmeric (Tyagi *et al.* 2004).

India is a rich repository of spices and over 100 species of spices are grown in about 2 million hectares in the country with an annual production of 2.02 million tonnes fetching Rs. 7,000 crores of revenue and Rs. 1184 crores worth of foreign exchange. It was the lure of Indian spices that attracted the ancient Egyptians, Romans, Arabs and the rest of the world to India. It was in search of spices, Vasco da Gama landed in the Malabar Coast of India which changed the course of Indian history. Spices have played a pivotal and colourful role in the history of mankind.

The latest list of ISO recognises 109 spices, but only a few are important commercially such as black pepper, cardamom, ginger, turmeric, large cardamom, cumin, coriander, fennel, fenugreek, garlic, saffron and celery. They acquire their characteristic odour from volatile constituents in the plant material. Spices are extensively used either directly or after extraction of volatile oils as condiments and flavouring materials. They are extensively used in flavouring food and confectionery, perfumery and soaps and also as colouring materials.

Many of these groups of crops have medicinal properties and are used in many Ayurvedic and Unani medicines. Inspite of the rich diversity and importance, the productivity of spices is low in India. However, the low productivity of various spices in India is because of various constraints. The more serious among them are the foot rot disease of black pepper caused by *Phytophthora capsici*, the 'Katte' disease of cardamom, the rhizome rot and bacterial wilt diseases of ginger etc. Non availability of disease free planting material is a major bottleneck in increasing the production and productivity of spices.

Tissue culture protocols for a large number of spices including black pepper, cardamom, ginger, turmeric, vanilla, large cardamom and many important herbal and seed spices have been developed. Spices and Aromatic crops being crops based on quality of the produce, it is essential to identify genotypes with high quality attributes and use the same genotypes for commercial planting. In this context, micro propagation protocols come in handy in multiplying the elite genotypes and production of disease free planting material. These techniques could also be used for multiplication of biomass and *in vitro* conservation of genetic resources of these crops. The plant regeneration protocols developed can be used for induction and exploitation of somaclonal variation for crop improvement especially in spices with limited variability in germplasm.

Tissue culture is the process of producing plants from tissue of the desired plant in an artificial nutrient medium under controlled environment. The plants so grown would be exactly similar to the mother plant in all aspects.

The science of plant tissue culture takes its roots from path breaking research in botany like discovery of cell followed by propounding of cell theory. In 1839, Schleiden and Schwann proposed that cell is the basic unit of organisms. They visualized that cell is capable of autonomy and therefore it should be possible for each cell if given an environment to regenerate into whole plant. Based on this premise, in 1902, a German physiologist, Gottlieb Haberlandt (father of plant tissue culture) developed the concept of *in vitro* cell culture. He isolated single fully differentiated individual plant cells from different plant species like palisade cells from leaves of *Laminum purpureum*, glandular hair of *Pulmonaria* and pith cells from petioles of *Eicchornia crassiples* etc and was first to culture them in Knop's salt solution enriched with glucose. In his cultures, cells increased in size, accumulated starch but failed to divide. Therefore, Haberlandt's prediction failed that the cultured plant cells could grow, divide and develop into embryo and then to whole plant. This potential of a cell is known as totipotency, a term coined by Steward in 1968. Keeping the above information under consideration the present study has been taken up with the following objectives.

a. Crop Improvement

As you all understand that for any crop improvement, conventional breeding methods are employed which involve six to seven generations of selfing and crossing- over to obtain a pure line. With plant tissue culture techniques, production of haploids through distant crosses or using pollen, anther or ovary culture, followed by chromosome doubling, reduces this time to two generations.

b. Micropropagation in largescale

Plant tissue culture techniques have also helped in large- scale production of plants through micropropagation or clonal propagation of plant species. Small amounts of tissue can be used to raise hundreds or thousands of plants in a continuous process. This is being utilized by industries in India for commercial production of mainly ornamental plants like orchids and fruit trees, e.g., banana. Using this method, millions of genetically identical plant can be obtained from a single bud. This method has, therefore, become an alternative to vegetative propagation. Shoot tip propagation is exploited intensively in horticulture and the nurseries for rapid clonal propagation of many dicots, monocots and gymnosperms.

c. Genetic Transformation

Tissue culture, in combination with genetic engineering is very useful in gene transfers. For example, the transfer of a useful bacterial gene say, cry (crystal protein) gene from *Bacillus thuringiensis*, into a plant cell and, ultimately, regeneration of whole plants containing and expressing this gene (transgenic plants) can be achieved.

d. Production of Pathogen-free Plants

Eradication of virus has been an outstanding contribution of tissue culture technology. It was found that even in infected plants the cells of meristem are either free of virus or carry an eligible amount of the pathogen. Such meristematic shoot tips are cultured in a suitable culture medium to obtain virus- free plants. This technique is economical and used very frequently in horticulture, production of virus- free ornamentals etc.

e. Production of Secondary Metabolites

Cultured plant cells are also known to produce biochemicals [secondary metabolites] like, alkaloids, terpenoids, phenyl propanoids etc. of interest. The technology of suspension cells is now available to the industry. The commercial production of 'shikonin' [a naphthoquinone] from cell cultures of *Lithospermum erythrorhizon*, has been particularly encouraging.

1. Protocols for Micropropagation of Black pepper (*Piper nigrum*)

Plant species: Black Pepper (*Piper nigrum* L.)

Family: Piperaceae

Black pepper is a universal table condiment used to flavor all types of cuisines worldwide. It is christened as the'King of Spices' (Srinivasan, 2007; Mathew *et al.*, 2001).

Origin and distribution: Black pepper is native to the tropical evergreen forests of Southwest India. It is cultivated commercially in India, Indonesia, Malaysia, Brazil, Sri Lanka and Vietnam.

Uses: Black pepper and white pepper are used mainly in flavouring of processed food, perfumery and in seasoning meat products. In medicine, it is used as an aromatic stimulant in cholera, as an antipyretic in malarial fever and arthritic diseases.The active principle in pepper is piperine, responsible for the biting taste. Black pepper is also an important traditional medicine and used to treat asthma, chronic indigestion, colon toxins, obesity, sinus, congestion, fever. It has been shown to have antimicrobial activity (Dorman & Deans, 2000).

Explants for micropropagation: Shoot tip and nodal segments for clonal propagation, stem and leaf tissues for plant regeneration and shoot tips for *in vitro* conservation.

Culture conditions and media: Surface sterilisation with 0.1% $HgCl_2$ for 5-10 min.

Incubation at 22±2° C with 14 h photoperiod of 3000 lux.

a) 1/2 WPM + 3 mgl^{-1} BA + 1 mgl^{-1} Kin for multiple shoots and callus regeneration.

b) 1/2 WPM without growth regulators for *in vitro* rooting.

c) 1/2 WPM + 15 g/l sucrose + 15 g/l Mannitol for *in vitro* conservation by slow growth.

The leaf explant from mature plant was cultured on MS medium supplemented with different concentrations of plant growth regulators (2, 4-D, BAP, IAA, Kinetin). Best callus was produced on MS medium with 4.0+3.0 mg/l 2, 4; D+ BAP by leaf explants (Dipali *et al.,* 2014).

***in vitro* responses:** Multiple shoots could be induced from shoot tips and nodal explants in 60 days and callus regeneration from leaf, shoot tip and stem explants was induced when cultured on 1/2 WPM + 3 mg/l BA and 1 mgl/I Kinetin within 90 days. Shoot tip-cultures could be maintained up to 1 year without subculture in 1/2 WPM + 15 g/l sucrose + 15 g/l mannitol.

Black pepper cultures are plagued by endogenous bacteria contaminating cultures. Use of antibiotics in the culture media help in reducing bacterial contamination. Tissue culture techniques have taken significant part in clonal propagation, conservation of germplasm and plant improvement in black-pepper (Sajc *et al*., 2000). That is why *in vitro* propagation is a superior option and alternate for traditional propagation.

Hardening and field performance: *in vitro* rooted plantlets were transferred to a mixture of garden soil, sand and vermiculite (1:1:1) with 80% success when kept in humid chamber for 20 days for hardening. Preliminary evaluation of micropropagated plants, based on vegetative characters after 2 years in the field, indicated that micropropagated plants are uniform and are similar to their parental stocks.

Applications: The protocols could be used for clonal multiplication of elite genotypes, creating somaclonal variation for crop improvement and for *in vitro* conservation and exchange of germplasm. Tissue culture techniques have taken significant part in clonal propagation, conservation of germplasm and plant improvement in black-pepper (Sajc *et al*., 2000).

2. Protocols for Micropropagation of Small cardamom *(Elettaria cardamomum* Maton.)

Plant species: Small cardamom *(Elettaria cardamomum* Maton.)

Family: Zingibcraceae

Origin and distribution: Cardamom grows wild in the rain forests of southern India and Sri Lanka at altitudes between 750-1500 m. It is now cultivated widely in India, Sri Lanka, Guatemala, Tanzania and Vietnam. In India, germplasm of cardamom (1,284 accessions) is maintained largely in the field genebanks (Ravindran *et al*., 2005).

Uses: Cardamom spice obtained by drying the fully developed fruit capsules is used on a world wide basis for domestic culinary purposes. The active principles in cardamom are á - terpenyl acetate and 1, S cineole. Tincture of cardamom is used in medicine for flatulence and stomach disorder. It is used as mouth refreshner and digestive aid. Biotechnological interventions would help in the introduction of genes conferring resistance to viral diseases and insect-pests into elite cardamom lines, thereby producing superior varieties. (Ravindran *et al*., 2000).

Explants for micropropagation: Rhizome bits with vegetative buds for micropropagation.

Culture conditions and media: Surface sterilisation with 0.1% $HgCl_2$ incubation at 22±2°C with 14 h photoperiod of 3000 lux

a) MS + 1 mg/l NAA for rooting.

b) MS + 1 mg/l BA + 0.5 mg/l NAA for multiplication and rooting.

c) MS without growth regulators for rooting

d) 1/2 MS + 15 g/l sucrose + 15 g/l mannitol for slow growth.

***in vitro* responses:** Multiple shoots up to S could be induced on MS with NAA within 90 days of culture and subsequently in every 70 days. Plants could be regenerated on MS + BA and NAA. Rooting was better in growth regulator free medium especially in liquid medium. Cardamom cultures could be maintained upto 1 year on V2 MS + 15 g/l sucrose + 15 g/l mannitol in sealed culture tubes. In vitro raised plantlets were established in soil with SO% success after hardening in a humid chamber for 30 days, in a mixture of garden soil, sand and vermiculite in equal proportions.

Hardening and field performance: After 5 years of field evaluation, tissue cultured plantlets performed similar to that of the control.

Applications: The protocols can be used to multiply elite genotypes, producing disease free planting materials and production of somaclones to exploit somaclonal variation for crop improvement. The slow growth method is ideal for *in vitro* conservation of germplasm.

3. Protocols for Micropropagation of Large cardamom*(Amomum subulatum* Roxb.)

Plant species: Large cardamom *(Amomum subulatum* Roxb.)

Family: Zingibcraceae

Origin and distribution: Large cardamom is native of Eastern Himalayas and is cultivated in India, Nepal, China and South East Asia. Large cardamom is native to the Sikkim Himalaya (Sharma *et al*., 2013).

The genus *Amomum subulatum* Roxburgh, (Family: *Zingiberaceae*, 2n=24 chromosome), commonly known as *"Bada Elaichi"* or "Black Cardamom" is a perennial herb, monocieous, and monocotyledon

Uses: The seeds are used in preparation of sweet meats. In medicine, they are fragrant adjuvants to other stimulants, bitter and purgative. The oil extracted from them is applied to eyelids to reduce inflammation. Cineole contributes to its pungency while terpenyl acetate towards the pleasant aroma. Large cardamom seed has a pleasant aromatic odour, due to which it is extensively used. for flavouring vegetables and many food preparations in India (Pathak, 2008).

Explants for micropropagation: Vegetative buds and rhizome bits with buds for micropropagation and vegetative buds for *in vitro* conservation Explants for tissue culture of the cultivars of *Ammomum subalatum* Roxb. *cv* Ramsey was collected from the nursery bed raised in the Sikkim State Council of Science and Technology, Rumtek. For preparation of explants standard sterilization protocol was followed. of (Pradhan *et al*., 2012)..

Culture conditions and media: Surface sterilisation with 0.1% $HgCI_2$ incubation at 22±2°C with 14 h photoperiod of 3000 lux.

a) MS + 1 mg/l BA + 0.5 mg/l IBA for micropropagation.

b) 1/2 M S + 10 g/lsucrose + 10 g/l mannitol for *in vitro* conservation by slow growth.

in vitro **responses:** Multiple shoots (5-10) could be induced from vegetative buds explants in MS with BA ($1 mgl^{-1}$) and IBA ($0.5\ mgl^{-1}$) within 60 days of culture. Plantlets could be maintained up to 1 year without subculture in 1/2 M S + 10 gl^{-1} sucrose + 15 gl^{-1} mannitol in screw capped culture tubes. The combination of cytokinins and auxins in the culture medium enhanced the response in a number of plants in terms of shoot and root growths, with 100% *in vitro* multipled plantlets were successfully transferred into field (Pradhan *et al.*, 2014).

Hardening and field performance: The rooted plants could be separated and transferred to soil with over 80% success. The field planted tissue cultured plants after 1 year, performed similar to· that of control.

Applications: Protocols can be used for multiplication of disease-free planting material of elite genotypes. The slow growth method is ideal for *in vitro* conservation of the species.

4. Protocols for Micropropagation of Ginger *(Zingiber officinale* Rosc.)

Plant species: Ginger *(Zingiber officinale* Rosc.)

Family: *Zingibcraceae*

Origin and distribution: Ginger is native of the tropical southern Asia and is now widely grown in India, China, Hawai, Japan, Pakistan, West Indies, Africa and Australia.

Ginger is an herb with white /yellow flowers, dark green leaves and a thick root. It is a medicinal plant used in ayurvedic medicine. Many products are manufactured from ginger i.e. dehydrated ginger, spices, ginger oil and oil resins etc.

Uses: It is used in sweet and savoury cooking, curry powder, spice blends and in medicine as a digestive aid. Ginger tea helps to improve the circulation and eases travel sickness. Gingerol and shogaol are the major constituents of ginger essential oil.

Explants for micropropagation: Vegetative buds, immature inflorescences and rhizome bits for clonal propagation. Vegetative buds, ovary, leaf and anther tissues for plant regeneration via callus phase and vegetative buds for *in vitro* conservation. Dwivedi *et al.* (2014) suggested that BAP containing medium is good for preserving different ginger varieties. The cultured varieties were able to micropropagate and showed very good bud initiation and shoot/root proliferation. These plants show 90% survival and regeneration. This technique is helpful for the preservation of ginger and could be a step towards preserving endangered plant species. Development of micro propagation protocols is a pre requisite for any successful *in vitro* conservation. (Sharma *et al.*, 2013)

Culture conditions and media: Surface sterilisation with 0.1% HgC12, incubation at 22±2°C with 14 h photoperiod at 3000 lux.

a) MS + 1 mg/l NAA for multiple shoots and *in vitro* rooting.

b) MS + 2 mg/l 2,4-D for callus induction.

c) MS + 10 mg/l BA + 0.2 mg/l 2,4-D for conversion of floral buds to plantlets and production of somatic embryos.

d) 1/2 MS + 10 g/l sucrose + 10 g/l mannitol for *in vitro* conservation by slow growth.

***in vitro* responses:** Ginger could be micropropagated from both vegetative buds and rhizome explants on MS medium + 1 mg/l NAA. Callus could be induced from different explants in MS + 2 mg/l 2, 4-D. Plant regeneration *via* callus derived from leaf, ovary, vegetative buds and anther was obtained on MS + 10 mg/l BA + 0.2 mg/l 2,4-D. Conversion of floral buds into vegetative buds was induced on the same medium. Ginger plantlets could be stored *in vitro* upto 1 year in 1/2 MS + 10 g/l sucrose + 10 g/l mannitol in screw capped tubes. An efficient and promising protocol for *in vitro* propagation of *Zingiber officinale* Rosco. using sprouting buds was established. Sprouting buds were sterilized and cultured onto MS medium supplemented with different growth regulators. Augmentation of MS-medium with 4.5 mg/l BAP recorded the highest percentage of shootlets multiplication. Shootlets were highly rooted on half strength of B5 medium supplemented with 1.0 mg/l NAA. The maximum percentage of acclimatization, hardening and rhizomes production of in vitro derived plants in greenhouse was 80–100 % (Abbas *et al.*, 2011). The application of BAP in the culture medium influences the rapid proliferation of meristems of many plants, including ginger. Some workers have reported the use of other cytokinins being beneficial for enhanced multiplication rates in ginger. (Rahman *et al.*, 2012 and Bhosale *et al.*, 2011). High sucrose (9%) in culture medium was supportive for induction of in vitro rhizomes in Zingiber officinale, and proved to be useful for its *in vitro* conservation (Tyagi *et al.*, 2006).

Hardening and field performance: In vitro rooted plantlets were established with 90% success, when transplanted into a mixture of garden soil: sand and vermiculite (1:1:1) and kept in humid chamber for 20 days. Field evaluation of tissue cultured plants indicated that it takes three crop seasons for the micropropagated plants to develop rhizomes of normal size; hence they cannot be used directly for commercial planting.

Applications: *in vitro* techniques can be used for in vitro pollination and seed set. Somaclonal variation can be an important source of variability to evolve high quality lines and lines resistant to rhizome rot. The slow growth method can be utilised for germplasm conservation. Somatic embryos encapsulated in sodium alginate, (synthetic seeds) are of immense use in germplasm exchange and conservation.

5. Protocols for Micropropagation of Turmeric *(Curcuma *longa* L. syn. *C. domestica* Val.).

Plant species: Turmeric *(Curcuma longa* L. syn. *C. domestica* Val.).

Family: Zingibcraceae

Origin and distribution: Turmeric is native to Indo Malayan region and is cultivated in India, Indonesia, China, Bangladesh, South America and Caribbean Islands.

Uses: Turmeric is used as a condiment and is an essential ingredient of curry powder for flavouring South Asian dishes. It is taken as a tonic, remedy for liver problems and to treat skin diseases. Curcumin from turmeric is also a traditional textile dye. In paste form, it is applied as a beauty mask and in face creams. Volatile oil consists of turmerones and tertiary alcohols.

Explants for micropropagation: Vegetative buds and rhizome explants with axillary buds for clonal propagation and vegetative buds for *in vitro* conservation.

Culture conditions and media: Surface sterilisation with 0.1% $HgCl_2$, incubation at 22±2° C with 14 h photoperiod at 3000 lux.

a) MS + 1 mg/l BA + 0.5 mg/l NAA for micropropagation and regeneration from callus.

b) 1/2 MS + 10 g/l sucrose + 10 g/l mannitol for slow growth.

Spontaneous rooting was observed after 4 weeks of cultivation with all treatments using half strength MS medium containing IBA and NAA at different concentrations, high frequency of rooting (89.76%) was obtained in 3.0 μM IBA containing medium. The plantlets, thus developed, were hardened and successfully established in natural soil. (Shahinozzaman *et al.,* 2013). The elongated shoots produced 9.66±1.20a roots on half strength MS liquid medium supplemented with 2 mg/l of IBA and showed 86 % survivability after hardening. Genetic fidelity of the micropropagated plantlets was confirmed using RAPD analysis employing 12 primers. This system provides high fidelity micropropagation system for efficient and rapid micropropagation of this important medicinal plant. (Ghosh *et al.*, 2013). Intraction between photoperiod condition and media types showed that dark condition with liquid and semisolid medium produced highest number of microrhizomes per shoot (7.00), followed by four hour light condition with liquid and semisolid medium (Chougule *et al.*, 2011).

***in vitro* responses:** Multiple shoots (8-10) could be induced from vegetative buds as well as rhizome explants of turmeric on MS medium with 1 mg/l BA and 0.5 mg/l NAA 40 days of culture. Production of callus and regeneration of plantlets from callus were also not iced in the same medium. Plantlets could be maintained up to one year without subcult ure in 1/2 MS + 10 g/l sucrose + 10 g/l mannitol in screw capped culture tubes. Best root formation of *in vitro* developed shoots could be achieved on halt strength MS medium supplemented with IBA at concentration 1.0mg/l. The *in vitro* developed plantlets were transferred to pot and they were grown in greenhouse for hardening and finally they were planted in the open field. Around 90% of plants were successfully established in natural field condition. (Gomathy *et al.*, 2014).

Hardening and field performance: The plantlets could be established in soil with over 80% success, when planted out in a mixture of garden soil, sand and vermiculite (1:1:1) and kept in humid chamber for about 20 days. It takes three crop seasons for micropropagated plants to produce normal sized rhizomes hence cannot be used directly for commercial planting.

Applications: Micropropagation techniques can be used for clonal multiplication of elite genotypes. Somaclonal v3.riation can be used for selecting better genotypes. The slow growth method can be used for *in vitro* conservation of germplasm. It was reported that high curcumin types were isolated from tissue cultured plantlets of turmeric.

Conclusion

Tissue cultured plants should be acclimatized before introduction to field and must be planted in the fields with minimum frost areas or else inside green house. Low multiplication rate in conventional propagation, micropropagation technology is the alternative method to supply healthy plant propagules continuously for commercial use. With regard to tissue culture of ginger, there are reports on successful micropropagation from rhizome buds. An optimum size of sterilized explants and suitable culture medium mainly require initiating *in vitro* cultures. *in vitro* culture techniques provide an alternative means of propagation and a tool for crop improvement. In addition, during storage and cultivation, rhizomes used for vegetative propagation are susceptible to diseases, that cause tissue senescence and degeneration. Heavy losses have been reported because of infection with soft rot (*Pythium aphanidermatum*) and nematodes (*Meloidogyne* spp.). Turmeric rhizomes degenerate after long term propagation and it is difficult to propagate through seed because of poor flowering and seed set. Thus In vitro culture techniques provide an alternative means of propagation and a tool for crop improvement.

References

Abbas, M. S., Hussein, S. T., Usama, I. A., Hattem, M. E. and Gaber, I. E. (2011). In vitro propagation of ginger (*Zingiber officinale* Rosco). *Journal of Genetic Engineering and Biotechnology*. 9(2): 165–172

Bhosale U.P. (2011). *In-vitro* shoots multiplication in different species of banana, *Asia. J. Plant Sci. Res.*, 1(3): 23-27

Chougule, P. S., Ramakrishna, V., Hegde, A. N., Mokashi, C. K., Venugopal, A. And Bhat, R. (2011). Microrhizome production in turmeric (*Curcuma longa* L.). *Karnataka J. Agric. Sci.* 24 (4): 493-496.

Dipali, M. and Wankhade, S. (2014). Investigation on callus formation in *Piper nigrum* L. *Int. J. Pharmaceutical and chemical science*. 3(2): 358-366.

Dorman, H.J. and S. Deans. (2000). Antimicrobial agents from plants; Antibacterial activity of plant volatile oils. *J. Appl. Microbiol.*, 88: 308- 316.

Dwivedi, H. S., Dwivedi, P. and Sharma, K. 2014. Preservation of Ginger (*Zingiber Officainale* Roscoe) through Tissue Culture Technique. *Journal of Environmental Research and Development*. 9(1): 192-195.

Ghosh, A., Chatterjee, P. and Ghosh, P. (2013). A protocol for rapid propagation of genetically *true to type* Indian turmeric (*Curcuma longa* L.) through *in vitro* culture technique. *Advances in Applied Science Research*. 4(3): 39-45.

Gomathy, V., Anbazhagan, M. and Arumugam, K. (2014). *In vitro* propagation of *Curcuma longa* (Turmeric). *International Journal of Research in Plant Science*. 4(1): 30-33.

Hussain, A., Naz, S., Nazir, H. and Shinwari, Z. K. (2011). Tissue Culture of Black Pepper (*Piper Nigrum* L.) in Pakistan. *Pak. J. Bot.* 43(2): 1069-1078.

Hussein Noreldaim, (2012). Effects of nutrient media constituents on growth and development of banana (*Musa* spp.) shoot tips cultured *in-vitro, Afri. J. Biotechnol.*, 11(37): 9001-9006.

Mathew, P.J., P.M. Mathew and V. Kumar. 2001. Graph clustering of *Piper nigram* L. (Black pepper). 118: 257-264.

Pathak A (2008), "Cultivation of Large Cardamom in Sikkim", Facets of the Northeast.

Pradhan S, Basistha BC and Handique PJ. 2012. "Determination of genetic fidelity of Micropropagated plants of *Zingiber officinale* cv-*Majhauley* of Sikkim Himalaya using RAPD markers", *Int. J. Fundamental Applied Sci.*, 1: 21-24.

Pradhan, S., Pradhan, S., Basistha, B. C. and Subba, K. B. (2014). *in Vitro* Micropropagation of *Amomum Subulatum* (*Zingiberaceae*), a Major Traditional Cash Crop of Sikkim Himalaya. *Int. J. LifeSc. Bt & Pharm.* 3(2): 169-180

Ravindran P. N.; Babu K. N.; Peter K. V.; Abraham Z.; Tyagi R. K. Spices. (2005). In: Dhillon B. S.; Tyagi R. K.; Saxena S.; Randhawa G. J. (eds) Plant genetic resources: horticultural crops. Narosa, New Delhi, pp 190–227

Ravindran, P.N. (2000). *Black pepper: Piper nigrum. Series: Medical and Aromatic Plants – Industrial Profiles.* Center for Medicinal Plants Research. Kerala, India. ISBN: 9057024535. Publisher Availability: In Stock CRC Press pp. 1-526.

Sajc, L., D. Grubisic and C. Vunjak-Novakovic. (2000). Bioreacters for plant engineering: An outlook for further research. *Biochem. Eng. J.,* 4: 89-99.

Sazedur Rahman (2013). Micro propagation of banana (*Musa* sp.) cv. Agnishwar by *invitro* shoot tip culture, *Int. Res. J. Biotechnol.,* 4(4): 83-88.

Shahinozzaman, M., Ferdous, M. M., Faruq, M. O., Azad, M. A. K. And Amin, M. N. (2013). Micropropagation of black turmeric (*Curcuma caesia* Roxb.) through *in vitro* culture of rhizome bud explants. *Journal of Central European Agriculture*. 14(3): 110-115.

Sharma. P. and Sarma K. P., (2013). *in vitro* propagation of *Bambusa tulda* : An important plant for better environment, *J. Environ. Res Develop.,* 7(3), 1216-1223.

Srinivasan, K, (2007). Black pepper and its pungent principle-piperine: A review of diverse physiological effects Critical Rev. *Food Nut.,* 47: 735-748.

Tyagi R. K.; Agrawal A.; Yusuf A. Conservation of Zingiber germplasm through *in vitro* rhizome formation. *Sci. Hortic*. 108: 210–219.

Tyagi R. K.; Yusuf A.; Dua P.; Agrawal A. (2004). *in vitro* plant regeneration and genotype conservation of eight wild species of Curcuma. *Biol. Plant*. 48: 129–132.

Section 03
Nursery Management

9

Modern Nursery Raising System in Vegetables

P. Pandiyaraj, S. Vijayakumar and Prashant K. Nimbolkar

Indian Agricultural Research Institute, New Delhi – 110 012, India
Indian Institute of Horicultural Research, Bengaluru–560 089, India

Introduction

Healthy seeds and seedlings are the first and essential requirement for achieving full yield potential of any vegetable crop. In recent years, vegetable growers have become highly conscious of the importance of quality seeds or seedlings. A major portion of the area under vegetable cultivation in India is now sown with hybrid seeds, which are costly but give higher yields and quality produce. In view of the high cost of seeds, some vegetable crops like tomato, brinjal, capsicum and cucurbits are being transplanted after growing nursery under protected conditions to achieve maximum germination count and healthy plant establishment. Raising of seedlings in plug trays or portrays is one such technology that achieves the above requirement. This technology is fast emerging as agro enterprise in India for its obvious advantage to both the nursery grower and the farmer. Soil-less substrate consisting of coco-peat and vermiculite is filled in plug trays and one seed is manually placed in each cell of the plug tray. However, manual seeding is labour-intensive operation requiring about more labour and time consuming. Mechanized precision seeding of plug trays is, therefore, necessary to enhance the capacity of rapidly growing nursery industry in India.

Importance of hi-tech vegetable nursery production

- Development of entrepreneurship in rural areas
- Development of related small scale input industries
- Employment generation especially for rural women
- Waste utilization – Coconut Coir pith

- Mechanization is possible, hence reduction in drudgery of farm workers
- Nursery men can also act as techno agent in spreading of technologies like improved varieties and hybrids
- Less loss of expensive seeds
- Proper seed germination, uniform growth, least seedling mortality
- Least pest and disease incidence

Table 1 : Yearly demand of vegetable seedlings and seeds in India

S.No	Crop	Seedling (millons)	Seed(kgs)
1.	Tomato	13028	600
2.	Brinjal	200000	635
3.	Chilli	14157	195
4.	Onion	695000	2779
5.	Cabbage	22963	101
6.	Cauliflower	12669	87

Source: *Resource Book on Horticulture Nursery Management, NAIP, ICAR, 2012*

Table 2 : Difference between filed grown seedlings and container grown seedlings

S.No	Traditional method of seedlings raising (Grown under open field condition)	Modern method of seedlings raisings(Grown in plug trays under protected condition)
1.	More seeds required for raising seedlings. For example hybrids seeds 100g per acre required.	Less seeds required for raising seedlings compared to traditional method. For example hybrids seed 70g per acre required.
2.	Disinfect the nursery area by solarization.	Use clean trays.
3.	Prepare seed beds about 3m X 1m in size and 20 cm heights.	Use the correct cell size according to the crop. For example, cucurbit crops require larger cell size than other vegetable crops.
4.	Prepare the nursery bed soil loose and friable and sow seed in lines about 5 cm apart and 1-2 cm deep. Adjust the planting depth according to the seed size.	Prepare growing media: normally coco peat mixes with bio-fertilizer were used. Sown one seed in each cell at a depth of 0.5 to 1.0 cm.
5.	Cover the nursery bed with paddy straw or dry leaves for germination.	Cover the trays with black polyethylene sheet for germination.
6.	Disinfect the nursery area by solarization.	Use clean trays.
7.	More loss of expensive seeds	Less loss of expensive seeds.

(Singh and Peter, 2014)

Components and Process Involved in Modern Nursery Raising System

Seedling tray, Media, Machanization, Irrigation, Nutrients, Protected structure, Light and Seed pelleting & priming, Biological enhancement and hardening.

Seedling tray

Choice of tray cell size (number and volume) to grow seedlings depends on several factors, including economics, plant growth rates, and customer demands. Cell size varies from 72 to 800 cells per standard tray (53.7 X 27.5 cm) (Hamrick, 2005). Cell size can vary by both the diameter (round cells) or size-length (square cells) and depth of the cell. Size than relates to cell volume, which controls the amount of media used as well as water holding capacity. It is less expensive on per plant basis to grow and ship a small volume cell than a larger one. The smaller the cell size dimensions in the tray, the more units a grower can produce per unit area since seedlings will finish faster and more seedlings can grow per unit area in the greenhouse space. Seedlings grown in larger cells are taller and have greater dry weight than those grown in small cells (Cantliffe 1993; Nesmith and Duval 1998). Tomato, brinjal, capsicum, chilli, watermelon, broccoli, cauliflower (Cantliffer 1993), cabbage (Marsh and Paul 1988), onion (Leskovar and Vavrina 1999) and muskmelon (Walters *et al.*, 2005) were reported to have earlier yields and more rapid early transplant growth when grown in larger cells. A container size increases root and shoot biomass is increase (Cantliffe 1993). If root volumes are reduce, water and nutrient uptake might likewise be reduced, thus leading to decreased shoot size.

Media

Media is one of the important component for modern nursery raising system. Successful nursery production is largely dependent on the chemical and physical properties of the growing media. Physical characteristics of the media are important for water and nutrient holding capacity as well as proper aeration for optimum root growth. Both physical and chemical components of media are most importance for optimum shoot and root growth. Choosing a media for seedling production also must be based on economic considerations. Costs for media may optimize plant growth, but costs for the products may not be economically viable. Cell volume then becomes as key economic factor with regard to the quantity of media required to grow the seedlings. Additionally, small cell volumes can create problems with moisture retention, aeration and soluble salt buildup. The type of irrigation method will also dictate media decisions.

A growing media must be sterile and may or may not be inert, but it should have cation exchange capacity (CEC). The CEC is associated with pH and nutrient buffering capacity (Agro 1997a). For this reason, many seedling production operations use peat as a major part of media makeup. On a weight basis, the CEC of sphagnum peat is about nine times that of a loamy mineral soil. However, due to the low bulk density of the peat, the effective CEC is lower than that of mineral soil. The ratio of water to air held in the media is controlled by particle size and pore space distribution (Argo 1997b). Capillary pores (< 0.3 mm) retain water after irrigation, while non-capillary pores (> 0.3 mm) provide root aeration (Argo 1998). Thus, it is important to provide an adequate mix of both capillary and non-capillary pore in the media.

Coco peat is a byproduct of extraction of fiber from coconut husk. It is 100 percent natural, biodegradable, fibrous and spongiest material. It has high C: N ratio, therefore it takes long period to decompose. Also it has high water holding capacity, approximately seven to nine times its own weight. Coco peat contains antifungal and antibacterial properties. In addition, it contains high CEC and moderately high pH value. Considering above properties, Coco peat is very good plant growth substrate, after adjusting several properties in soil less cultivation such as green house vegetable nurseries.

Seed pelleting and priming

Many vegetable seeds that are sown for seedling are small or big and irregular shaped. Most commercial seedlings producers use drum vacuum seeders thus, these types of seeds should be pelleted or coated. Pelleting builds up seed size and uniformity by layering the seed with layers of clay type material and binder. Seeds normally pelleted include tomato, brinjal, chilli, onion, lettuce and celery. Seed coating is accomplished by applying wet solutions containing a dye and many times fungicides or other chemical treatments. The coating does not build up seed size, nor does it change the shape of the seed. The coating makes the seeds more flowable in the seeders. Species that are commonly coated include cucumber, watermelon, squash and melon. Using dyes in the coating or pelleting process allows the seeds to be more visible at sowing (Cantliffe 2009).

Biological enhancement

Seed Treatment is a most common method adopted for all types of inoculants. The seed treatment is effective and economic. Treat the seeds with Trichoderma viride 4 g or Pseudomonas fluorescens 10 g or Carbendazim 2 g per kg of seeds 24 hours before sowing. Just before sowing, treat the seeds with Azospirillum @ 40 g / 400 g of seeds. Mix sterilized cocopeat @ 300 kg with 5 kg neem cake along with Azospirillum and Phosphobacteria each @ 1 kg. Approximately 1.2 kg of cocopeat is required for filling one protray.

Machanization

The transplant tray is the first area that facilitates mechanization both in seedling and potentially automatic field transplanting (Shaw 1993). Essentially all seedling transplant operations have mechanized the seeding process (Styer *et al*., 1997). Trays are loaded in bulk onto a line where they are filled with media, dibbled, sown via a vacuum drum seeder, topped with vermiculite, watered and then stacked by hand. Operators have to inspect the line, stack trays to fill, and stack the filled flats as they come off the line. This process is enhanced if the seeds are pelleted, round and of reasonable size. More sophisticated seedling operations use robots to load and unload flats onto benches.

Germination

Warm temperature and uniform moisture are needed to ensure successful germination and get the plants off to an even start. Many germination chamber systems are commercially available including custom built germination units. Many growers use bottom heat or root zone heating to provide warm, even temperature. A weed mat (Black polyethylene) is placed on the top of the bench to help spread the heat with skirts on the side to help contain the heat. Tomato seeds germinate are best at 21^{0}C. The ideal root zone temperature is 26 to 29^{0}C during the first four weeks of growth and 20 to 26^{0}C during the fifth and sixth weeks. In brinjal seed germinate at 21 to 24^{0}C. Chilli seed germinate at 28 to 32^{0}C (Singh and peter 2014).

Irrigation

Seedlings can be irrigated above the plants or sub irrigation via ebb and flow. The costs vary considerably. Over head irrigation usually requires a boom with nozzles, while the more expensive ebb and flow system requires concrete floors and a recycling system for the water. The advantage of the latter system is that no water is applied to the foliage of the seedling. Therefore, there is reduced potential for development and spread of diseases. Also, water and nutrients are recycled, reducing groundwater pollution, thus saving water and fertilizer. Although the type of irrigation system may alter root morphology in the seedlings, total yields of many vegetable crops are not affected by the type of irrigation (Franco and Leskovar 2002). Flotation systems improved uniformity of tomato and bell pepper (Leskovar 1998). Total root elongation was similar for both overhead irrigation and sub irrigation treatment. Sub irrigation reduced shoot to root ratio and promoted plant hardiness.

Nutrients

Nutrition plays an important role in the nursery. The nutrition in the form of organic and inorganic sources supplies all the essential nutrients required for proper growth and development of the nursery plants. The availability of nutrients to the plant depends upon many factors such as source material, nutrient content, correct soil condition and type of fertilizer. Plant performance in nursery depends on nutrient status in plant media, but no nutrition guideline has been worked out owing to its short duration of plants with nursery. It is necessary to apply need based nutrients for quality production of seedlings. Both shoot and root mass were improved by increasing N, but the increase in shoot mass was much greater than for root mass in response to N, resulting in higher shoot to root ratios. Generally, when adequate rates of P are supplied to the growing media, little enhancement of transplant growth and performance in the field is observed when P levels are increased (Soundy *et al.*, 2001).

Apart from the N, P, K, Ca, Mg and S, micronutrients such as Fe, Mn, Zn, Cu, B, Mo and Cl are also necessary for maintaining the normal health and metabolism of plants. Nutrient deficiency leads to poor and stunted growth of plants and resulting in poor performance. Nutrient deficiency can easily be corrected by applying the deficient element either by soil application or by foliar application or by the combined application of soil application and foliar spray (Ganeshamurthy and Rupa, 2016).

Protected structure

Poly house gives better protection, due to total avoidance of rain water entry into poly house, hence leaf diseases can be easily controlled. Transparent UV stabilized polyethylene film 200 micron thickness is used for covering the poly house roof. It is provided with retractable or movable shade nets, at about 11 feet height just below the structures from ground level. The sides of the poly house are covered with 200 micron thick polyethylene film to a height of 3 feet from the ground level, to have better protection from rain splash. Remaining height of side wall is covered with 40 micron white coloured insect proof net from all the four sides.

Poly house is provided with an ante chamber with two doors constructed at opposite directions where entry or exit to the poly house is made through the first door and then after closing the first door, the second door is opened to make an entry into the poly house. Care should be taken not to open both the doors simultaneously to avoid the entry of pests into the protected structures. A small concrete trough of 2 meter length, 1 meter breadth and 2 inch depth should be prepared between the two doors of antechamber for facilitating washing legs in the disinfectant solution (Potassium permanganate) to prevent any contamination inside the poly house (Hebbar *et al.*, 2011).

Economics of raising nursery under poly house

Size of structure (10m (L) x 5m (W) x 7' (H))	= 50m^2
No. of seedlings 8000 (polybags) + 7000 (portrays)	= 15,000/-
Seedling @ Rs.2/-	= Rs. 30,000/-
Investment cost (polyhouse, polybags, protrays)	= 15,500 + 5,000
	= Rs. 20,500/-
Net profit 1st year (30,000 – 20,500)	= Rs.9500/-
Net profit 2nd year (30,000 – 5000)	= Rs.25000/-
Net profit 3rd year (30,000 – 6000)	= Rs.24, 000/-

(Yadav *et al*., 2014)

Light

Light is one of the important environmental factors that control seedlings growth. Light may also be used to condition seedlings before they are transplanted to more hostile environments. New photo selective greenhouse films now allow growers to alter light quality inexpensively. Improving light quality can improve the seedlings growth. Under a fixed photoperiod, seedling growth and quality was improved by increasing the photosynthetic photon flux (PPF) (Kitaya *et al.* 1998). PPF was adjusted for all film treatments. The FR light absorbing films were effective in reducing stem elongation and producing compact tomato, bell pepper and cucumber seedlings.

Hardening

The term, hardening off, refers to the process of gradually exposing seedlings that have been raised in protective covering system to the climatic condition they will be grown in. Depending on vegetable plant variety, 4-6 weeks after germination, the seedlings should well grown and ready to plant out into the garden, before this can be done, they must first be hardened off. This is done to reduce stress and subsequent growth check when seedlings are transplanted to the main garden. Hardening can be accomplished by increasing light intensity or exposing transplants under full sunlight, reducing irrigation or watering and fertilizer application

Pest and disease management

- Hygiene has a vital role in the control of pests and diseases.
- Use steam or chemical sterilization of the growing media, structures, tools and trays.

- Effective ventilation and air movement is also a sound disease prevention method.
- Understand pests and diseases that could affect the growth of healthy seedlings
- Care must be taken with the use of pesticides within enclosed areas.
- Note also that in a greenhouse plants can be more sensitive to chemicals than in the open field.

Limitations

- Lack of trained personnel and skilled labors for doing nursery activities
- Initial establishment and maintenance cost for nursery needs to be met by the people, which they presume as a risk bearing activity
- Risk on marketing of seedlings, pest and disease damage incidences
- Transplant shock which delays growth but is not as severe on cell raised seedlings compared to bare rooted seedlings.

Conclusion

Vegetable nursery production has become a highly commercialized business, wherein most farmers buy their plugs from professional growers. Many factors help to produce quality seedlings. These include the use of high quality seeds, growing media with good drainage, water holding capacity, and providing optimal rates of fertility. Further, seedlings are germinated under more or less optimal conditions to obtain uniform stands and are grown in protected culture under greenhouse conditions. Rate of plant development, root structure, plant height and vegetative matter can be tightly controlled under these conditions. Container cell sizes can be adjusted to help produce plant sizes that conform to strict can be adjusted to help produce plant sizes that conform to strict customer demands. The use of robots for tray filling, in the production greenhouse, and the mechanization of both the planting and growing process, as well as fertilization of the tray and harvest process can further reduce labor requirements of the plug production system. The advent of plug transplants have allowed growers of many specialty crops the ability to greatly reduce seed costs, increase stand uniformity, and in many cases increase yields and quality of the products produced. In the future, many more crops may be grown as seedling transplants, especially those of high economic value and potentially high seed cost.

References

Argo, W. R. (1997a). Transplant production and performance: root medium chemical property. 5th Natl. Symp. Stand Establishment. *Transplant Proc.* 4-10.

Agro, W. R. (1997b). Transplant production and performance: media physical properties. In: 5th Natl. Symposium on Stand Establishment. *Transplant Proc.* 11-14.

Argo, W. R. (1998). Root medium chemical properties. *HortTechnology* 8: 486-494.

Cantliffe, D. J. (1993). Pre and postharvest practices of improved vegetable transplant quality. *HortTechnology*, 3: 415-418.

Cantliffe J. Daniel. (2009). Plug transplant Tecghnology. *Horticultural Reviews,* 35: 397-436.

Franco, J. A. and Leskovar, D. I. (2002). Root dynamics of muskmelon transplants as affected by nursery irrigation. *J. Am. Soc. Hort. Sci.,* 127: 337-342.

Ganeshamurthy, A. N. and Rupa, T. R. (2016). Nutrient management in nurseries. Compendium of lectures, Indian Institute of Horticultural Research.

Hamrick, D. (2005). Ornamental bedding in plant industry and plug production. Pp. 27-38. In: M. B. McDonald and F. Y. Kwong (eds.), Seeds: Biology and technology. CAB International. Oxfordshire, UK.

Kitaya, Y., Niu, G., Kozai, T. and Ohashi, M. (1998). Photosynthetic photo flux, photoperiod, and CO_2 concentration affect growth and morphology of lettuce plug transplants. *Hort Science,* 33: 988-991.

Leskovar, D. I. (1998). Root and shoot modification by irrigation. *Hort Technology,* 8: 510-514.

Leskovar, D. I. and Vavrina, C. S. (1999). Onion growth and yield are influenced by transplant tray cell size and age. *Scientia Hort.* 80: 133-143.

Marsh, D. B. and Paul, K. B. (1988). Influence of container type and cell size on cabbage transplant development and field performance. *Hort Science,* 23: 141-142.

NeSmith, D. S. and Duval, J. R. (1998). The effect of container size. *Hort Technology* 8: 495-498.

Shankara S Hebber. (2011). Protected cultivation of capsicum. Technical Bulletin No. 22. Indian Institute of Horticultural Research.

Shaw, L. N. (1993). Changes needed to facilitate automatic field transplanting. *Hort Technology* 3: 418-420.

Singh, D.K. and Peter, K.V. (2014). Protected cultivation of horticultural crops. New India Publishing Agency. Pp. 223-258.

Soundy, P., Cantliffe, D.J., Hochmuth, G.J. and Stoffella, P.J. (2001). Nutrient requirements for lettuce transplants using a floatation irrigation system. I. Phosphorus. *Hort Science* 36: 1066-1070.

Styer, R.C. and Koranski, D.S. (1997). Plug and transplant production: a grower's guide. Bell Publ., Batavia, IL

Walter, S.A., Riddle, H.A. and Schmidt, M.E. (2005). Container cell volume and transplant age influences muskmelon development and yield. *J. Veg. Sci.* 11: 47-55.

Yadav, R.K., Kalia, P., Choudhary, H. and Zakir Husainand Brihama Dev. (2014). Low-cost poluhouse technologies for higher income and nutritional security. *International. J. of Agri. and Food Sci. Tech.* 5: 191-196.

Section 04
Rootstock Management

10

Rootstock Management of Fruit Crops

Trina Ahikary

Department of Fruit Science, Punjab Agriculture University, Ludhina, Punjab

Introduction

Controlled perpetuation of plants is known as plant propagation. Plant propagation can be broadly classified in two types: sexual and asexual. In sexual propagation, the fusion of male and female gametes gives rise to a new plant, whereas in asexual propagation which is also known as vegetative propagation, the development of new plants takes place without the fusion of male and female gametes, that is, the new plant develops from vegetative plant parts. Various techniques of asexual propagation are adopted in plant propagation, such as: grafting, budding, layering, cutting, etc. So far cutting and layering are concerned; the new plants are developed on their own root system, but in grafting and budding on the others' root system. It means that in these two techniques (grafting and budding), two plant parts are joined to form new plants; the one which forms the root system is known as rootstock and the one forming the shoot system is known as scion. Plant propagation is both a science and an art. The science of plant propagation requires knowledge of plant physiology, nursery cultural practices, and characteristics of the particular plant that you want to grow. The art of plant propagation cannot be taught in a book or classroom, however, because it consists of specific technical skills that must be acquired through innate ability or experience and often requires a certain feel.

The interactions (relationships) between these two plant parts (rootstock and scion) are of much significance in the commercial fruit growing industry. According to Hartmann *et al.*, 2002 combining two different plants (genotypes) into one plant by grafting or budding can produce growth patterns that are different from those that would have occurred if each component part had been grown separately. Some of these effects are of major importance in horticulture and, while others are deimental and should be avoided. These altered characteristics may result from (a) specific characteristics of the graft partners not found in the other, for example, resistant to certain diseases, insects or nematodes – or tolerance to certain adverse environmental and soil conditions; (b) interactions between the rootstock and the scion that alter size, growth, productivity, fruit quality or other horticultural attributes; and (c) incompatibility reactions.

In practice, it may be difficult to separate which of the three kinds of influencing factors dominant in any given graft combination growing in a particular environment. Long-term results depend on the rootstock scion combination, environment (climate, edaphic factors such as soil), propagation, and production management; this affects yield, quality and plant form and the economics of production.

Thus, rootstocks provide growers with a useful tool to manipulate the vigour and performance of orchard trees. Effects on tree size, precocity, fruit production and maturity are achieved through complex interrelationships between the roots and canopy of the plant. Rootstocks directly affect the ability of plants to take up water and nutrients. They are also able to significantly alter the pattern of canopy development and functions such as photosynthesis.

Types of Rootstocks

Rootstocks are of different types depending on their multiplication methods or raising methods. They are:

Seedling rootstocks: Seedling rootstocks are those developed from germinated seeds, have certain advantages. Production of seedlings is relatively simple and economical and is well adapted to mass propagation methods. Most seedlings plants do not retain viruses occurring in the parent plants (although some viruses are seed transmitted). In some instances (e.g. plum rootstocks), the root system developed by seedlings tends to grow deeper and to be more firmly anchored than rootstocks grown from cuttings. However, seedling rootstocks have disadvantage of genetic variation, which may lead to variability in the growth and performance of the scion of the grafted plants. Such variation is most likely to occur if the seed is obtained from unknown, unselected sources. The seed source plants may be unusually heterozygous or may have been cross pollinated with related species. Within the same species, some seed sources produce much better rootstock plants than others. With careful testing, individual plants in a species could be selected to be a mother tree seed source, or the start of the clonal rootstock. Variability among seedling rootstocks can be reduced by careful selection of the parental seed source regarding its identity, and by its protection from cross pollination. Variation can be reduced if all nursery trees of the same age are dug from the nursery row at one time, and small or obviously off type seedlings or budded trees are discarded. In most nurseries, the young trees are graded by size; all those of the same grade are sold together. The practice of retaining slow growing seedlings or budded trees for an additional year's growth is undesirable as it tends to perpetuate variability. Many fruit orchards, if grown on uniform seedling rootstocks, show no more variability resulting from the rootstocks used than from unavoidable environmental differences in the orchard, principally soil variability (Hartmann *et al.*, 2002).

Clonal rootstocks: Rootstocks, which are propagated vegetatively, are called clonal rootstocks. Each clonal individual plant is genetically the same as all the other plants in the clone and can be expected to have identical growth characteristics in a given environment (Hartmann *et al.*, 2002). Propagation by vegetative means avoids genetic variations resulting in true to type progenies. In carrying out budding and grafting, two plant parts are required namely rootstock and scion. However, using, clonal rootstocks in budding and grafting, multifarious effects can be observed/obtained due to interaction of rootstocks and scion cultivars. Clonal rootstocks can be raised from micro-propagation in apple, apricot, almonds, chestnuts, cherries etc. Clonal rootstocks are desirable not only to produce uniformity but equally important to preserve their special characteristics and the specific influences they have on scion cultivars, such as disease resistance, dwarfing characteristics, growth, or flowering habit.

Characteristics of Good Rootstocks

Once the rootstocks are used and orchard is raised by heavy investment, it is not possible to change without incurring serious losses (Rajput and Haribabu, 1995). As rootstocks have its effect on many aspects of plant growth and development, they are generally used to combat market and field problems like tree life, adverse conditions (soil, disease, and climate), fruit quality, and crop maturity, yield and so on. For every rootstock to be utilized in propagation, it should possess the following characteristics:

- Should exhibit a high degree of compatibility with scion cultivars and give maximum productive life to the trees.
- Should be well adapted to the agro-climatic conditions of the particular locality like frost, cold and heat.
- Should be resistant to diseases and pests prevalent in the concerned area.
- Should be tolerant to adverse soil conditions like salt and drought.
- Must exhibit favourable and positive influence on the performance, bearing and quality of scion variety.
- Should possess good nursery characteristics like germination, high degree of polyembryony, ability to attend graft-able size in a short period and free from excessive branching.

Effect of Stock on Scion Cultivates

Size and Growth Habit

In apple, rootstock can be classified as dwarf, semi dwarf vigorous and very vigorous rootstock based on their effect on a scion cultivator. If a scion is drafted on dwarf rootstock the graft combination will be dwarf while he same cultivar grafted on very rootstock would grow very vigorously. In citrus, trifoliate oranges. On the other hands, in mango all plants of a given variety are known to have the same characteristic canopy shape of the variety despite the rootstocks being of seedling origin. But recently, rootstock of kakarady, olour have been found to impart dwarfness in the scion cultivators. Guava cultivars grafted on *psidiumpuminum* are found to be dwarf in statue.

Precocity in Flowering and Fruiting

The time taken from plating to fruiting precocity is influenced by rootstocks. Generally fruiting is influenced by rootstocks. Generally fruiting precocity is associated with dwarfing rootstocks and slowness to start fruiting with vigorous rootstocks. Mandarins, when grafted on Jamberi rootstock are precious than those grafted on sweet orange or sour orange or acid lime rootstock. Some invigorating rootstocks, such as M.25, induce precocious flowering and cropping in apple.

Floral Abundance

There are several ways in which rootstocks can affect floral abundance. First, they may alter scion habit in favour of the production of many short shoots which terminate in floral buds (dards or brindles); this is of particular value with tip-bearing cultivars of apples suchas 'Cortland' or 'Worcester Pearmain'. Second, they may increase the proportion of axillary buds on 1 -year-old wood which develop florally rather than vegetatively. Whilst this may increase the potential for fruit set on young wood, it often results in this wood becoming bare of buds (blind) in subsequent seasons ; this can be a particular problem with some sour (tart) cherry (*Prunuscerasus* L.) cultivars. Some rootstocks may also increase the production of floral spurs on 2-year or older wood. Finally, rootstocks may influence the number of floral buds/spur or the number of flowers in each floral bud.

Fruit Set and Yield

The rootstock directly influence on the production of flowers and setting fruit in oriental persimmon (*Diospyrous kaki* cv. Hichiya). When it is grafted on D. Lotus I products more flowers but few only mature but when *D. Kaki*is used as the rootstock, the set is more.

The influenced of rootstock on the yield performance or cultivar has been well documented in many fruit crops. Acid limes budded on rough lemon register nearly 70 percent increased yield than those budded on troyercitrange, Rampur lime or its own rootstock. Sweet orange var. Sathgudi budded on Kichili rootstock gave higher yield than on Jamberi or on its own seeding (South India).

Fruit Size and Quality

Sathgudi sweet oranges grafted on gajanima rootstock produced large but poor quality fruit, while on its own roots they produced fruit with high juice content and quality. The physiological disorder 'granulation' in sweet orange is very low of grafted on Cleopatra mandarin seedling, on their hand rough lemon seedling, stocks induced maximum granulation. The physiological disorder 'black end' in Barlett Pear did not appear if *PyrusCommunis* was used as the rootstock when *P. pyrisfolia*was used as the rootstock. This disorder appeared, affecting fruit quality.

Nutrient Status of Scion

Rootstocks do influence the nutrient status of scion also. Sathgudi sweet oranges trees have a better nutrient in the leaves when on it's budded on *C. volkarminriana* rootstock than on its own rootstock or Cleopatra mandarin stocks.

Winter Hardiness

Young grape fruit tree on Rangpur lime withstand winter injury better than on rough lemon or sour on orange. Sweet orange and Mandarins on trifoliate were more cold hardy.*C. unshiu* tolerant to freezing condition.

Diseases Resistance

In citrus considerable variability exists among the rootstock in their response to diseases and nematodes. For instance rough lemon rootstock is tolerate to tristeasa, xyloprosis and exocortisis tolerant to gummosis but susceptible to exocortisvirus disease. Similarly guava varieties grafted on chinse Guava, resist wilt diseases and nematodes.

Ability to Resist Soil Adverse Conditions

Among the citrus rootstocks trifoliate orange exhibits poor ability to resist excess soil moisture or excess boron in the soil. Myroblam plum rootstock generally viz, peach, appreciator almond.

Frost sensitivity

The simplest way in which rootstocks influence the frost sensitivity of trees is by advancing or delaying their time of flowering. Delayed flowering may decrease, whereas advanced flowering may increase, the risk of frost damage. However, there is also some evidence that rootstocks may, in addition, influence the resistance of individual flowers to frost damage. For example, Durner (1990) has reported that the blossoms of peach (*Primus persica* (L.) Batsch) cultivars grown on the French rootstock GF.677 are more sensitive to frost damage than blossoms from trees on other rootstocks.

Effect of Scion on Root Stock

Effect on Root System of Stock

In apple it has been found that if apple seedling were budded with the "Red Astrochan "apple the rootstock produced a very fibrous root system with few top roots. On the other hand, if scion cultivar is less vigorous than the rootstock cultivar the rate growth and the dry ultimate size of the tree is more determinate by the scion rather the rootstock.

Cold Hardiness of the Rootstock

Cold hardiness of citrus roots is affected by the scion cultivar. Sour orange seedling budded to 'Eureka 'lemon suffered much more from winter injury than the unbudded seedlings.

Age of Root Stock Seedling

Young mango rootstock seedling (6 month to one year old) were found to put forth inflorescence when the branches from old trees are inarched which will be attributed to the influence of scion on the rootstock.

Various Factors are Influence on the Successful Graft or Bud Union Those are as Follows

Incompatibility

Certain rootstock and scions are incompatible; therefore, the graft union between these two will not normally take place.

Kind of Plant

Some species like oats are difficult to graft but apple and pears are very easy in predicting a successful graft union.

Environmental Factor During and Following Grafting

There are certain environment al requirements which must be met for callus tissue to develop and heal the graft union.

a) Temperature has a pronounced effect on the production of callus tissues. An optimum temperature as essential for the production of callus tissue. In most of the temperature fruit crops callus production is retarded after 100 0 F.

b) Relative humidity must be high or maintaining a film of water against the callusing surface is essential to prevent these delicate thin walled parenchymatous calls from drying.

c) Presence of high Oxygen content near this surface is essential.

Growth Activity of Stock Plants

Some propagation methods such as "T" budding and bark grafting depend upon the bark grafting depend upon the bark 'spipping' which means the cambila cells activity dividing and producing young thin walled cells on the side of the cambium. These newly formed cells separating readily from one another as the book slips.

Propagation Techniques

Sometimes the technique used in grafting is so poor that only a small portion of the causal regions of the stock and scion are brought together. This measurement in its failure of the graft union.

These major effects of rootstocks are being exploited in a beneficial manner in the fruit industry. However, the knowledge of specific rootstock effects on specific scion cultivars is of utmost importance to get maximum benefits from the enterprises.The various parts of the tree maintain a high degree of autonomy in fruit crops. Different varieties grafted on the same tree often leaf out and bloom at different times, although they are supported by the same root system. It would not be realistic to expect a rootstock to induce radical change in fruit quality, a 'Valencia' orange will always have the characteristics of this variety no matter on what rootstock it is grown. On the other hand, the roots are the absorbing organ for mineral elements and water and growth regulators are formed in them. The statement that compounds not normally found in the scion cannot be introduced by a rootstock is generally still true, although later work showed that organic compounds characteristic of the rootstock sometimes can be translocated into the top of the tree.Rootstocks and scion interact in many ways including at least 14 fruit quality factors influenced by the rootstock. A high-quality variety, such as 'Valencia,' has adequate quality on any rootstock, but marginal varieties may not. Fruit destined for the fresh market must have good internal quality, but the appearance of the fruit (size, shape, peel colour, peel thickness, interior colour, seediness, and granulation) is also important. In fruit for processing, soluble solids production per acre is the most important criterion, but with the great number of 'Hamlin' orange trees now in production, juice colour has become a concern of the processors and there is interest in improving the juice colour of early orange varieties through rootstocks. Yield, soil adaptation, and disease resistance are still the key factors in selecting rootstocks, especially for fruit grown for processing. There are signs, however, that more attention will be given to solids content and juice colour. It is not clear how rootstocks exert their influence on fruit quality, some of the rootstock effects appear to be size-related (larger fruit have lower solids or based on nutritional effects (potassium increasing fruit size and acidity). It has to be kept in mind that rootstock effects sometimes vary from year to year, from area to area, and with cultural practices.

Identification and utilization of appropriate rootstocks in crops like apple, pear, citrus, mango and grapes in relation to vigour management, nutrient uptake, soil salinity, moisturestress and yield efficiency has been already studied in several countries. Rootstocks with immense potential for commercial exploitation are yet to be utilized to its full potential most of the commercial fruit crops in India. Therefore, it is important step to find a rootstock with the maximum desirable characteristics to meet the set of environmental conditions where the tree is to be grown. It means a root-stock considered best for a certain variety and environmental conditions may not be a best choice for some other variety and environmental regime.

Rootstock Effects in Different Fruit Crops

Role of RootStocks in Mango

Mango is one of the most important tropical fruit of the world which is known as 'king of fruits' in India. Some attempts have been made to standardize the rootstocks for various scion varieties including the use of polyembryonic varieties for vigour management, salinity and drought tolerance as well as to improve fruit yield and quality. Vigour management plays an important role in mango, especially for high density planting and orchard management in terms of canopy management, harvesting and plant protection measures. According to Jauhari *et al.*,1972, growth and bearing habit of Dashehari mango on the seedling stock of Dashehari itself, showed most vigorous nature and highest yield compare to other polyembryonic rootstocks. Similar results were reported from IIHR-Bangalore, based on 21 years study on the performance of Alphonso which indicated nucellar seedlings of Muvandan, Bappakai and Olour were vigorous rootstocks in decreasing order of vigour while Vellaikulumbanseedlingimparted dwarfing in comparison to Alphonsografted on its own seedling (Reddy *et al.*, 2003). Some factors like phenolic contents bark percentage and chlorophyll fractions were found to be very useful for determining the vigour of mango rootstocks at nursery stage (Abirami *et al.*,2011). Murti *et al.*, 2003 suggested that selection of size controlling rootstocks at the nursery stage can be made by assessing leaf water potential, total phenols/and or ABA. Fruit yield and productivity plays an important role in crops like mango where the productivity is very low in most of mango growing countries. Smith *et al.*, 2008 reported that yield and yield efficiency of 'Kensington Pride' was best on Sg. Siput rootstock and it was poor with Sabre rootstock indicated the possibilities for manipulating mango scion productivity through rootstock genotypes. Quijada *et al.*, 2008 found that Sinamaica rootstock have the best adaptability to agro-ecological conditions of Maracaibo plain of Venezuela and produce higher yield and production efficiency with respect to different cultivars (Criollode Mara, Manzana and Sensation mango). Bappakai was found to be best rootstock for Dashehari followed by Muvandan and EC 95862(Chandan *et al.*,2006). Langra grafted on Bappakai rootstock recorded the highest fruits number/plant followed by Vellaikulumban and Chandrakaran Singh *et al.*, 2006. Fruit quality also important aspect and some studies were done on this aspect in different countries. Rootstock trial conducted at IIHR, Bangalore for 21 years, indicated no significant effect of rootstocks (Vellaikulumban, Bappakai, Chandrakaran, Kurukan, Muvandan, Mylepelian and Olour) on fruit quality of Alphonso mango (Reddy *et al.*, 2003). Screening of mango rootstocks to salinity has shown that the polyembryonic cultivars Olour and Bappakai could withstand higher level of salinity (Palaniappan *et al.*, 2001). Dubey *et al.*, 2007 reported that Olour as best salt tolerant rootstock compared to Kurukan. Mango cultivar 13-1 was selected as a polyembryonic (3-6 embryos) rootstock for calcareous soils or for irrigation with saline water. Mango trees on 13-1 rootstocks showed excellent performance on soil containing 20% lime, three other cultivars on 13-1 rootstock showed good development on sandy soil with 10-20% lime Gazit *et al.*,1980. ZuazoDurán *et al.*, 2004 indicated the greater capability of 'Gomera-1' rootstock to saline conditions. Seedlings from stone of Kesar variety was found to be better with significantly highest survival percentage, germination percentage and growth parameters with higher salt concentrations of water (Varu and Bharad, 2010).

Role of Rootstock in Citrus

Role of rootstocks in citrus is one of the most debatable and discussed issues and its selection is a major consideration under planning of any citrus orchard. One thing can be safely said that choosing the right rootstocks is fundamental to the success of the orchard. The citrus rootstock scenario in India has been reviewed by Patil *et al.*,1987. The characteristics of some selected rootstocks are given in Table 1.

Table 1: Characteristics of some selected citrus rootstocks (Chadda and Singh, 1990)

Rootstock	Horticulturalperformance					Reaction to				
	Yied	Quality	Plating material vigour	Root	Citrus nematode	Tristeza	Exo-cortis	Salt	Drought	Root system
Rangpur lime	G	M	G	M	S	R	S	R	R	D
Marmalade orange	G	M	G	MT	MT	R	S	R	R	D
Rough lemon	G	L	G	S	S	R	R	T	T	D
Cleopata mandarin	M	M	M	T	S	R	R	M	S	M
Sour orange	G	G	M	R	T	HS	T	T	MT	D
Sweet lime	M	M	G	S	MT	S	S	S	S	M
Trifoliate orange	L	M	L	R	R	R	H	HS	HS	SH
Troyercitrange	M	G	M	MT	T	MT	S	HS	HS	SH
Carrizo citrange	M	G	M	MT	T	MT	S	HS	HS	SH
Sweetorange	G	G	M	HS	HS	MT	R	S	S	M
Karnakhatta	G	M	M	S	MT	-	T	T	S	D
Nasnaran	M	G	M	S	MT	T	R	HT	S	M

G = good, M = Moderate, L = Poor or low, R = Resistant, T = Tolerant, MT = Moderately tolerant, S = Susceptible, HS = Highly susceptible, D = Deep, M = Medium, SH = Shallow, - = No information

Role of Rootstock in Grapes

Although grapevines can grow in soils with a wide range of pH (4.5-6.5), very acid soils present a problem. Based on vine growth, Himelrick, 1991 found different responses of *Vitis* species or cultivars to low soil pH. The cultivars most tolerant to strong acid soils were *V. labrusca* cv. 'Concord' and 'Catawba', along with rootstock 'SO_4' and'3309C', and the hybrid cultivar 'Seyval'; *V. vinifera* 'White Riesling' and 'Chardonnay' were the most intolerant. The use of acid-tolerant rootstocks, such as 'SO4' and '3309C', was highly recommended. According to Leon, 1969 grape rootstocks *V. champini* and *V. vinifera* are considered to be tolerant tosalinity. Walker *et al.,* 2000 studied different grape rootstocks for certain level of salinity (0.4 or 2.3 dS/m) and concluded that grape cv. Shiraz grafted on 'Ramsey','1103 Paulsen' and '140 Ru' had higher wine K+ , pH and color hue rather than grafted on its own seedling. Same researchers also noted that grape cv. 'Sultana' grows vigorously when grafted on 'Ramsey', '1103P' and 'R_2', rootstocks. On the basis of nature of tolerance to different salt concentrations (0, 50, 85, 120, 155 mMNaCI),Troncoso *et al.,*1999 categorized different rootstocks as a sensitive (41 B, R. Lot, 110 R, 140 R and 161-49), moderately tolerant (13-5 and Ramsey) and tolerant (196-17, CH-1, CH-2 and Superior). Ezzahouani *et al.,*1995 suggested that hybrids

('110R', '140Ru' and '1103P') from rootstocks *V. berlandieri x V. rupestris* can be used in drought prone areas where water is a limiting factor for grapevine productivity and he also noted that drought resistance classification of rootstocks might vary from country to country. SimilarlyKocsis *et al.*, 1999 suggested that hybrids of *V. berlandieri x V. riparia* were more tolerant to drought.

There have been only a few studies related to rootstock effects on scion to cold hardiness. Miller *et al.*, 1988 found differences of cane and bud hardiness of rootstock 'K5BB', '3309C' and 'SO4'. Rootstock '3309C' had the most cold hardy canes and buds; its acclimation in fall was faster and declamation in spring was slower than 'K5BB' and 'SO4'. Striegler *et al,* 1991found that the rootstock did not affect the distribution of hardy canes within the canopy; 'Seyval' grafted on '3309C' appeared to be the cold hardy.

Role of Rootstocks in Guava

At present, guava is propagated on seedlings raised from open pollinated seeds and not on uniform clonal rootstocks. Edwar and shanar 1964 reported that *P. cattleianum, P. guinesee, P. molle* and Philippine guava were found suitable as rootstocks. On *P. cattleianum*, the trees were the tallest and gave the maximum yield. *P. pumilum* rootstock had dwarfing effect but fruits on this rootstock had maximum number of seeds, highest TSS and total sugars. Trees on *P. cujavillis* produced the largest fruits with the highest ascorbic acid content, though they were rough-skinned and not uniform. All rootstocks were free from wilt disease (*Fusariumsolani /Macrophominaphaseoli*) except 'Allahabad Safeda' Teaotia and Pandey1961. Sharma *et al.,* 1992 recorded that rootstock aneuploid No. 82 impart dwarfness to 'Allahabad Safeda' in terms of plant height, plant spread and tree volume. They also recorded higher yield on this rootstock.

Role of Rootstocks in Apple

There is a great diversity in the type of material used for raising rootstock seedlings in apple. In Kashmir, wild indigenous 'Crab apple' known as 'Trel' is used, in Himachal Pradesh, seedlings of 'Crab C' are employed as rootstock Nauriyal, 1965. The first rootstock trial of apple was initiated in 1937 at Chaubattia with 'Red Delicious', 'Jonathan' and 'Rymer' cultivars using Crab C, M 2, M 13, Merton 779, Merton 793 and 'Local Selection' (seedling selection from 'Ribbistin Pippin') rootstocks. After the introduction of M and MM series of rootstocks during early sixties, from East Malling Research Station, England, at Mashobra and Kotkhai in HP, Chaubattia in UP and Shalimar in J & K, elaborate varietalcum-rootstock trials were initiated at Mashobra in 1967, Kotkhai in 1968, Chaubattia in 1969 and several locations in J & K in 1969. As a result, some promising rootstocks were identified for different apple growing regions in India viz. M 7, M 9, M 26, MM 106 and MM 111 for Himachal Pradesh; M 2, M 4, M 7, M 9 for J&K and Merton 779, MM 106, M 13 for UttarakhandAnanda, 1993. At Chaubattia, Om and Pathak, 1983 observed that M 2 rootstock response to prolific bearing and high yield in 'Red Delicious', 'Jonathan' and 'Rymer' at earlier stages (5th and 10th years). Subsequently, the trend changed later years (20th and 25th years) which recorded maximum yields in 'Merton 779' and 'Crab C' rootstocks whereas Merton 779 and M 13 rootstocks produced higher yields than other rootstocks after 35th year. Similarly, Sharma and Chouhan 1990 observed that apple trees grafted on MM 106 rootstock had significantly higher fruit set, yield and yield efficiency than those on M 7. Chandal and Chouhan 1992 observed that 'Starking Delicious' on MM 111, MM 106, MM 104 and M 4 rootstocks showed less reduction in growth, photosynthetic efficiency and nutrient uptake, and had higher stomatal resistance, lower transpiration rate, more accumulation of proline, ABA and carbohydrate under water stress (10 bar) and these rootstocks were thus considered more drought tolerant than M 2, M 7, M 9, M 25, M 26 and MM 109.

Role of Rootstocks in Pear

Fewer rootstocks choices are available for pears than for apples. Domestic pear seedlings (*Pyruscommunis*) are still the most acceptable rootstocks for pear cultivars in terms of vigour, hardiness, and compatibility. However, all pear trees on seedling roots are susceptible to fire blight. Seedlings of *Pyruscalleryana* are adapted to many soil conditions and produce semi-vigorous growth. It is resistant to fire blight, but it's not sufficiently winter hardy. *Pyrusbetulaefolia* seedlings are also adapted for many soil conditions and it's a winter hardy. The trees are vigorous, larger than 'Bartlett seedling', and moderately tolerant to fire blight. *Pyrusussurensis* can be used as resistant to fire blight, pear psylla, and cold hardy.

Role of Rootstocks in Stone Fruits

In India, the stone fruits like peach, plum, apricot and almond are generally propagated on their own seedling, while the cherry plants are raised on 'Paja' (*Prunuscerasoides*) seedlings. Commonly 'Behmi (*P. mira*)' is widely used as a rootstock for almond (Chadda and Kakara 1977). The relative contribution of organic and inorganic solutes to osmotic adjustment in three almond rootstocks subjected to soil salinity showed that leaf water and osmotic potentials were affected by salinity in GF677 and Bitter almond, but less so in GN15, suggesting a higher selectivity for K+ and Ca2+against Na+ in this latter rootstock. Peach seedlings itself use as a rootstocks for peach plantation and plum, apricot, almond seedlings also used (Dewan, 1963 and Singh and Guptha 1971). Peach seedlings generally shows susceptibility to nematode but 'Nemaguard' and 'Okinawa' showed tolerance to nematode attack (Hasen *et al.,* 1956). Some nurseries also use 'Behmi' (*P. mira*) as rootstock for peach. Apricot can be used as resistant rootstock against root knot nematode. Western sand cherry, Almond, Nanking cherry are dwarfing rootstock for peach. For Plum (*P. domestica*) selections like 'Brompton' and 'Common plum' from; St. Julien, 'Common Mussel' and 'Damson' from *P. institia*; Myrobalan from *P. cerasifera;* peach (*P. persica*), apricot (*P. armeniaca*), almond (*P. amygdalus*) and some hybrids like Marianna plum (*P. cerasifera* × *P. munsoniana*) are used as rootstocksin other countries. In India, peach, apricot, behmi (natural hybrid of almond and wild peach) and plum seedlings are commonly used as rootstocks (Singh and Guptha 1971). The main cherry rootstocks in other countries are Hazard (*P. valium*) and Mahler (*P. mahaleb*) seedlings, while Stockton Morello (*P. cerasus)* is also occasionally used. In India, the most common rootstock in Kashmir, Himachal Pradesh and Kumaon Hills of UP is 'Paja' (*P. cerasoides*) (Srivasthava and Gusthav 1963), though old plantations are invariably on Mazzard or Mahaleb rootstocks (Singh and Guptha 1971). For exceptional hardiness the seedlings of 'Russian sour cherries' could be used for *Prunusbesseyi* (Singh and Guptha 1971).

Conclusions

Rootstock and interstock effects on the vigour of growth, flowering, fruit set, yield efficiency and fruit quality of deciduous fruit tree scions are many, complex, and poorly understood. A better understanding of the mechanisms underlying these effects would aid future rootstock breeding and selection. Further research is required, particularly on the possible involvement of endogenous hormones in rootstock effects on scion growth and cropping. Comparisons of rootstocks budded with scions and growing in the orchard often yield quite different conclusions when conducted on different sites or with different scion cultivars. A better understanding of these interactions, betweenrootstocks and scions, soil, or climatic factors would aid more efficient selection and use of rootstocks in the future. All of this will only be achieved by greater collaboration between researchers working on rootstock projects in different parts of the world.

References

Abirami, K. Singh, R. Baskaran, V. (2011) Studies on the influence of seedling physiological parameters with vigour in some polyembryonic and monoembryonic mango genotypes. *Indian. J. Hortic,* 68: pp. 18-23.

Ananda, S.S. (1993) Rootstocks of stone fruits. *Advances in Horticulture: Fruit Crops.* Malhotra Publishing House, New Delhi, India.

Chadha, K. L. and Singh, H. P. (1990). *Citriculture scenario of India. in Citriculture in North-western India:* Proceedings of Citrus Show-cum-Seminar. Prospects and Problems of Kinnow Cultivation (K.S. Gill, J.S. Kanwar and R. Singh, eds.). Punjab Agricultural University, Ludhiana pp. 21-64.

Chadha, T. R. and Karkara, B. K. (1977) Morphological and anatomical characteristics of some indigenous Prunus species and their horticultural significance. In: *Fruit Breeding in India (Nijjar, G S., Ed.).* Oxford and IBH Publishing Co., New Delhi, 1977, pp. 122-129.

Chandal, J. S. and Chauhan, J.S. (1992) Drought Resistant of Starking Delicious apple plants on different Rootstocks. *J. Ind. Hort.,* 49(4): pp.294-299.

Chandan, P.M., Kadam, J.H. and Ambad, S.N. (2006) Effect of different polyembryonic and monoembryonic rootstocks on performance of Dashehari mango. *Inst. J. Agril. Sci.,* 2(2): 594-595.

Dewan, S.D. (1962). Peach cultivation in Punjab. *Hort. J.* pp. 96102.

Dubey, A.K., Srivastav, M., Sharma, Y.K, Pandey, R.N., Deshmukh, P.S. (2007). Dry mass production and distribution of nutrients in two mango rootstocks as affected by salinity. *Indian. J. Hortic.,* 64:385-390.

Durner, E. F. (1990). Rootstock influence on flower bud hardiness and yield of Redhaven peach. *Hort. Science,* 25:172-173.

Edwar, J. C. and Shankar, G. (1964). *Rootstock trial for guava. Allahabad Farmer,* 38:249-50.

Ezzahouani, A. and Williams, L.E. (1995). Influence of rootstock on leaf water potential, yield, and berry composition of Ruby Seedless grapevines. *Amer. J. Enol. Viticult.* 46:559-563.

Gazit, S. and Kadman, A. (1980). 13-1 mango rootstock selection. *Journal of Horticultural Science,* 15(5): 669.

Hansen, C.J., Lownsbery, B.F. and Hesse, C.O. (1956). Nematode resistance in peaches. *California Agric.* 10(9): pp. 5-11

Hartmann, H.T., D.E. Kester, F.T. Davies Jr., and R.L. Geneve. (2002). *Plant Propagation Principle and Practices* (6th edn.).Prentice Hall of India Pvt. Ltd. New Delhi.

Himelrick, D.G. 1991. Growth and nutritional responses of nine grape cultivars to low soil pH. *Hort. Sci.* 26: 269-271.

Jauhari, O.S., Teaotia, S.S. and Upadhyay, S.K. (1972). *Acta Horti.,* 24:107-109.

Kocsis, L., Granett, J., Walker, M.A., Lin, H., Omer, A.D. (1999). Grape phylloxera populations adapted to Vitisberlandieri x V. riparia rootstocks. *Am.J.Enol.Vitic.* 50:101-106.

Leon, B., Ehlig, C. F. and Clark, R. A. (1969). Effects of grape rootstocks on chloride accumulation in leaves. *J.Amer.Soc.Hort.Sci.,* 94:584-590.

Miller, D.P., Howell, G.S. and Striegler, R.K. (1988). Cane and bud hardiness of own-rooted White Riesling and scions of White Riesling and Chardonnay grafted to selected rootstocks. *Amer. J. Enol. Viticult.* 39: pp. 60-66.

Murti, G.S.R., Upreti, K.K. (2003). Endogenous hormones and phenols in rootstock seedlings of mango cultivars and their relationship with seedling vigour. *Eur.J.Hortic.Sci,*68: 2-7.

Nauriyal, J.P. 1(955). Apple in Punjab. *Farm Bulletin No. 10.* Punjab Agricultural University, Ludhiana, Punjab

Om, H. and Pathak, R. K. (1983). Influences of rootstocks on the growth and productivity of some apple cultivars. *J. Ind. Hort.,* 40(4): pp.168-172.

Palaniappan, Germplasm screening for salinity stress in tropical fruit species. Regional Training Course "Characterization, Evaluation and Conservation of Tropical Fruits Genetic Resources", organized by IPGRI, ICAR and IIHR2001.

Patil, V. K. (1987). High density planting and dwarfing rootstocks in citrus. A review. *J. Maharashtra Agric. Univ.* 12:189-94.

Quijada, O., Castellano, G., Jiménez, N. and Briceño, E. (2008). Development and productivity of mango cultivars grafted over sinamaicarootstcks in the plain of Maracaibo. *Agronomía Trop.* 58(3): 207-213.

Rajput, C. B. S. and R. S. Haribabu. (1995). Citriculture. Kalyani Publisher, New Delhi, pp. 147.

Reddy, Y. T. N., Reju, M., Kurian, P. R., Ramachander, Gorakh Singh and Kohli, R. R. 2003. Long term effects of rootstocks on growth, fruit yielding patterns of Alphonso mango (*M. Indica*) *Scientia Hort.,* 97(2): 95-108.

Sharma, D. D. and Chauhan, J.S. (1990). Effect of different rootstocks and training system on growth and cropping of 'Delicious apples. *Ind. J. Hort.,* 74(4): 365-370.

Sharma, Y.K., Goswami, A.M. and Sharma, R.R. (1992). Effect of dwarfing aneuploid guava rootstock in high density orcharding. *Indian J. Hort.,* 49: 31-36.

Singh, M. and Kanpure, R. N. (2006). Performance of mango cv. Langra on different rootstock. *Intl. J. Agric. Sci.,* 2(2): 564-565.

Singh, R. N. and Guptha, P.N. (1971). Rootstock problem in stone fruits and potentialities of wild species found in India. *Punjab Hort. J.* 11:157-175.

Smith, M. W., Bright, J.D., Hoult, M.D., Renfree, R.A. Maddenrn, T. and Coombes, N. (2008). Field evaluation of 64 rootstocksfor growth and yield of 'Kensington Pride' mango. *Hort. Science,* 43:1720-1725.

Srivastava, D. N. and Gustav, E. F. (1962). Propagation of temperate fruit plants in Kumaon hills. In fruit Nursery Practices in India. Directorate of Extension, Ministry of Food and Agriculture, New Delhi, pp. 14-18.

Striegler, R. K. and Howell, G. S. (1991). The influence of rootstock on the cold hardiness of Seyval grapevines I. Primary and secondary effects on growth, canopy development, yield, fruit quality and cold hardiness. *Vitis.* 30:1-10.

Teaotia, S. S. and Pandey, I. C. (1961). Effect of growth substances on the rooting of guava stem cutting. *Sic. Cult.*27: pp. 442-44.

Troncoso, A., Matte, C., Cantos, M. and Lavee. S. (1999). *Evaluation of salt tolerance of in vitro grown grapevine rootstock varieties. Vitis.* 38 (2):55-60.

Varu, D. K. and Barad, A.V. (2010). Standardization of mango rootstock for mitigating salt stress. *Indian J. Hort.,* 67: 79-83.

Walker, R.R., Read, P.E. and Blackmore, D.H. (2000). Rootstock and salinity effects on rates of berry maturation, ion accumulation and colour development in Shiraz grapes. *Austral. J. Grape and Wine Res.,* 6: 227-239.

Zuazo Durán, V. H., Martínez-Raya, A., Aguilar Ruiz , A. and Franco Tarifa, D. (2004). Impact of salinity on macro- and micronutrient uptake in mango (*Mangiferaindica*L. cv. Osteen) with different rootstocks.

